AF885797

PAPERS AND PROCEEDINGS
OF
APPLIED GEOGRAPHY CONFERENCES

Volume 16, 1993

Toronto, Ontario
Canada

Editors: J. W. Frazier, State University of New York - Binghamton
B. J. Epstein, Kent State University
F. A. Schoolmaster, III, University of North Texas
K. G. Jones, Ryerson Polytechnical Institute
N. J. Siller, Ryerson Polytechnical Institute

Editorial Assistant: Michael Kazmierczak

ISSN 0747-5160

APPLIED GEOGRAPHY CONFERENCE
STEERING COMMITTEE, 1992-93

Chair: John W. Frazier, SUNY - Binghamton
Co-Chair: Bart J. Epstein, Kent State University

MEMBERS AND AFFILIATIONS

B. Foust
University of Wisconsin - Eau Claire
Eau Claire, Wisconsin

R. Gordon
Rural Health Office
Phoenix, Arizona

J. Haake
Fay's Drug Company, Incorporated
Syracuse, New York

Col. R. C. Ham
United States Military Academy
West Point, New York

N. F. Henry
SUNY - Binghamton
Binghamton, New York

E. Hillsman
Oak Ridge National Laboratory
Martin Marietta Corporation
Oak Ridge, Tennessee

K. G. Jones
Ryerson Polytechnical Institute
Toronto, Ontario, Canada

P. D. Koos
Van W. Emmons, Sr. Population
and Marketing Analysis Center
Wysox, Pennsylvania

J. Lee
Kent State University
Kent, Ohio

J. D. Lord
University of N Carolina
Charlotte, North Carolina

J. F. Lounsbury
Arizona State University
Tempe, Arizona

Henry Moon
University of Toledo
Toledo, Ohio

J. Morrison
U. S. G. S.
Reston, Virgina

F. A. Schoolmaster, III
University of North Texas
Denton, Texas

K. Smith
May Stores
St. Louis, Missouri

J. Sobel
U.S. Census
Suitland, Maryland

L. Sommers
Michigan State University
East Lansing, Michigan

N. Torrieri
U.S. Census
Suitland, Maryland

APPLIED GEOGRAPHY CONFERENCE

TABLE OF CONTENTS

16TH ANNUAL

APPLIED GEOGRAPHY CONFERENCE

October 13-16, 1993

SPONSORED BY

SUNY - Binghamton
Kent State University

COSPONSORED BY

Ryerson Polytecnical Institute

CONFERENCE DIRECTORS

J. W. Frazier
SUNY - Binghamton
607-777-2755

B. J. Epstein
Kent State University
216-672-2062

K. G. Jones
Ryerson Polytecnical Institute
416-979-5038

F. A. Schoolmaster
University of North Texas
817-565-2091

LOCAL DIRECTOR

N. J. Siller
Ryerson Polytecnical Institute
416-979-5000 Ext.6177

HOST DEPARTMENT

School of Applied Geography
Ryerson Polytecnical Institute
350 Victoria St., Toronto, Ontario, M5B 2K3
416-979-5038

LOCATION

All Sessions take place at:
The Westbury Hotel
475 Yonge Street
Toronto, Ontario, M4Y 1X7
416-924-1413

AIR CANADA is the Official Airline for the Toronto Conference. **Save up to 50%**, pending availability, with **minimum guaranteed savings of 15%** on the full Hospitality and Executive Class services.

Please note, that, in Canada and the United States, **Air Canada offers an additional 5% reduction on its Excursion Fares**, with the exception of seat sale fares. **UNITED AIRLINES** is the **AIR CANADA** partner in the U.S.

To take advantage of the above savings, please call **your travel agency or Air Canada 1-800-361-7585**. When purchasing your ticket, please ask that the Event Number **CV930244** be entered in the Tour Code box and Reference Code **RPI** in the Endorsement box, **regardless of the fare purchased.**

CONFERENCE REGISTRATION FEES

Registration fees are $70.00 Cdn. before September 15, $80.00 Cdn. after September 15, and $30.00 for students.

SESSION CLASSIFICATION

A = Retailing

B = Environment

C = Economic Development

D = GIS Applications

E = Education

F = Military Geography

THEME SESSIONS

I: North American Retail Development

II: Water Resource Issues in the Great Lakes Basin

III: NAFTA

IV: Ontario Waste Management Issues

V: Toronto Waterfront Regeneration

WEDNESDAY, October 13, 1993

Wednesday Evening

8:30-10:30 p.m.

HOST RECEPTION - The Westbury Hotel

THURSDAY, October 14, 1993

9:00

PLENARY SESSION

M. Doucet, Ryerson Polytecnical Institute
Toronto: World Class City?

9:30-12:30p.m.

THEME SESSION I: NORTH AMERICAN RETAIL DEVELOPMENT

Organizer and Chair:
J. Konarski, International Council of Shopping Centers, New York

A. McKichan, Retail Council of Canada, Toronto
The Canadian Retail System of the 1990's

J. Bullock, Cadillac-Fairview Corporation Ltd., Toronto
The Canadian Shopping Centre System: Characteristics and Future Prospects

G. Brown, Woolworth Corporation, New York
Towards an Integrated North American Retail -Myth or Realty?

PANEL DISCUSSION:

K. Smith, The May Co., St. Louis
G. Kerfoot, Homart, Chicago
L. Carlson, Dayton Hudson Corp., Minneapolis
C. Ray, JMB Retail Properties Co., Chicago

CONCLUDING REMARKS:

K. G. Jones, Centre for the Study of Commercial Activity
Ryerson Polytecnical Institute

9:15-10:45 a.m.

SESSION B1

WATER RESOURCES I

Organizer and Chair:
J. A. Harrington Jr., Indiana State University

R. L. Repic, R. Hill-Rowley, Michelle Nichols and **Ken Koleda**, University of Michigan-Flint
The Development of Regional Groundwater Databases for Michigan's "Thumb"

R. A. Earl and **R. B. Anaya**, Southwest Texas State University
Managing and Modelling a Dilemma: Springflow from San Marcos Springs and the Endangered Species Act

V. P. Gutowski, Eastern Illinois University
Benefits and Problems in Using a Side-Channel Reservoir Configuration, Charleston, Illinois

S. M. Berta and **Timothy Gress**, Indiana State University
SCS Wetland Recertification Using SPOT Data

J. A. Harrington Jr. and **Lisa M. B. Harrington**, Indiana State University
Modelling the Expansion of Spartina in the Palix Estuary

SESSION B2

ENVIRONMENTAL POLICY I

Organizers and Chairs:
B. Coates, Lakeland Community College
J. Lounsbury, Arizona State University

F. Himmler, University of North Alabama
Groundwater Protection Policy: Alabama, Mississippi, and Tennessee Compared

G. Hepner, University of Utah
Environmental Effects of Land Use Change on the US-Mexico Border

C. Renata, Ontario Hydro
T.B.A.

SESSION C1

ECONOMIC DEVELOPMENT PLANNING I

Organizer and Chair:
P. D. Koos, PMAC

PANEL DISCUSSION:

P. D. Koos, PMAC
E. D. Jardine, PMAC
G. Marcho, PMAC
B. Edgcomb, PMAC

SESSION D1

INNOVATIVE USE OF GIS

Organizer and Chair:
D. Banting, Ryerson Polytecnical Institute and **C. Lo**, Gartner Lee

B. Foust, University of Wisconsin-Eau-Claire and **H. Botts**, University of Wisconsin-Whitewater
Street/Road Intersection Density and Population: Unintended Uses of the 1990 TIGER Files

C. Lo, Gartner Lee
T.B.A.

G. Doucet, North York Hydro
T.B.A.

D. Barron, Meridian Mapping Service
T.B.A.

11:00-12:30

SESSION B3

WATER RESOURCES II

Organizer and Chair:
F. A. Schoolmaster, University of North Texas

O. W. Templer, Texas Tech University, **L. V. Urban**, Texas Tech University and **D. Hawkins**, City of Lubbock Water Utilities
Municipal Water Supply Resources forLubbock, Texas: An Ongoing Problem

J. M. Harlin, Northern Illinois University
Impact of Agricultural Contaminants on Major Rivers in the Cornbelt

O. P. Matthews, Oklahoma State University
ASCE Model State Water Code

F. A. Schoolmaster, M. Acevedo, and **R. Hanes**, University of North Texas
Estimating the Impact of a Possible Free Trade Agreement on the Water Resources of Venezuela

S. Arbona, Shippensburg University, and **J. M. Hunter**, Michigan State University
Land Use and Water Quality in Puerto Rico

SESSION C2

ECONOMIC DEVELOPMENT PLANNING II

Organizer and Chair:
B. Warf, Kent State University

J. Cox, Canadian Office of Federal-Provincial Relations, and **V. Preston**, York University
Financial Services and Toronto: A Case Study in Corporate and Spatial Concentrations

J. A. McIntyre, Economic Development Co.
Work Force Quality's Influence on Business Location Decisions

B. Holly, Kent State University
Economic Development Implications of Changes in Commercial Banking: A U.S. Example

M. Welsh, Kent State University
Information Technology and Global Restructuring: Perspectives on Regional Development

B. Warf, Kent State University
The Pentagon and the Rustbelt

SESSION C3

ECONOMIC DEVELOPMENT PLANNING III
Organizer and Chair:
P. D. Koos, PMAC

PANEL DISCUSSION: STATUS REPORT ON THE BEEP PROJECT

L. W. Cooley, PMAC
A. J. O'Connor, PMAC
J. R. Dungan, PMAC
L. A. Gallagher, PMAC

SESSION B4

ENVIRONMENTAL POLICY II: CITIZEN INVOLVEMENT

Organizer and Chair:
M. Bardecki, Ryerson Polytecnical Institute

D. Stonehouse, City of Toronto Planning and Development Department
Public Involvement in Bringing Back the Don

S. Lerner, University of Waterloo
T.B.A.

M. Bardecki and **D. Banting**, Ryerson Polytecnical Institute
Citizen Environmental Advisory Committees: Involvement and Influence

SESSION D2

GIS I

Organizer and Chair:
J. Lee, Kent State University

J. Lein, Ohio University
Qualitative Reasoning & Environmental Decision Making: The Role of Logical Models

D. Sui, Texas A & M University
GIS-Based Shift-Share Analysis: An Application in Metropolitan Economic Modeling

E. K. Mbobi, Kent State University
A Multispectral Examination of the Buckeye Basin Wetlands

L. Marion, Kent State University
Effect of Data Aggregation on Computing Slopes and Aspects with Digital Elevation Models

J. Lee and A. Boros, Kent State University
Analyzing Information of Network Flows within a Geographical Information System

12:30-2:00p.m.

LUNCH

GUEST SPEAKER:

A. C. Lea, Compusearch Micromarketing Data and Systems, Toronto

2:30-5:30

THEME SESSION II - WATER RESOURCE ISSUES IN THE GREAT LAKES BASIN

Organizer and Chair:
M. Sanderson, University of Waterloo
Water Resource Issues in the Great Lakes Basin

F. Quinn, Great Lakes Environmental Laboratory
The Great Lakes Hydrology Explained

D. Cuthbert, Canada Centre for Inland Waters, Burlington, Ontario
The International Joint Commission Reference on Fluctuating Lake Levels

L. Mortsch, Atmospheric Environment Service, Toronto
Climate Change and the Great Lakes

PANEL DISCUSSION:

"IS WATER INCLUDED IN NAFTA? OR WILL THERE BE GREAT LAKES DIVERSIONS IN THE FUTURE?"

Organizer and Chair:
B. Mitchell, University of Waterloo
PANEL:

M. Clark, Ottawa
F. Quinn, Environment Canada, Ottawa
T. Kierans, Grand Canal Project, Newfoundland
J. Jackson, Great Lakes United

2:15-3:45 p.m.

SESSION A1

INTERNATIONAL RETAILING

Organizer and Chair
R. Muncaster, Wilfrid Laurier University

L. Sparks, Institute for Retail Studies, University of Stirling, Scotland
Spatial Restructuring of Food Retailing in Europe

A. Hallsworth, University of Portsmouth, England
Commercial Structural Changes in Britain as a Result of Land-Use and Regulation Changes

J. Simmons, University of Toronto
Comparing Commercial Structure: Hong Kong and North America

E. Kulke, University of Hanover, Germany
Spatial Effects of the Enterprise Concentration Process in German Retailing

SESSION C4

TOURISM AND RECREATION

Organizer and Chair:
J. Burke, Ryerson Polytecnical Institute

D. R. Crowe, Western Illinois University
Riverboat Casino Competition: An Issue in Sustainable Tourism

G. Nelson, University of Waterloo
T.B.A.

T. Manning, Centre for a Sustainable Future
Planning Tourism Development in SensitiveEnvironments

SESSION D3

RETAIL, GIS APPLICATIONS

Organizer and Chair:
M. Williams, R. L. Polk and Company (Canada)

L. Crabtree, R. L. Polk and Company (USA)
Development of GIS Software Applications in the Retail and Automotive Sectors

K. Allen, R. L. Polk and Company (Canada)
A Comparative Assessment of Approaches to Market Analysis in the US and Canada

M. D'Andreta, R. L. Polk and Company (USA)
T.B.A.

SESSION C5

HEALTH AND THE ENVIRONMENT I

Organizer and Chair:
E. Carlson, Ryerson Polytecnical Institute

M. Taylor and **J. Eyles**, MacMaster University

Psychosocial Effects of Environmental Contaminants: The Case of Smithville, Ontario

A. Baker, University of Toronto
Seasonal Affective Disorders

T. Sly, Ryerson Polytecnical Institute
Shifts in Disease Location in Global Warming

4:00-5:30 p.m.

SESSION A2

RETAIL PLANNING I

Organizer and Chair:
J. S. Cirillo, J. C. Penney Co.

R. Davis, Weingarten Realty Investors
Implications of NAFTA on Texas Retailing

W. D. Laufenbert, J. C. Penney Co.
Retail Market Analysis in U.S.-Mexico Border Markets

J. L. Rice, Holliday, Fenoglio & Co.
Houston's Grocery-Anchored Retail Centers: Geographical Perspectives Within the Real Estate Cycle

J. S. Cirillo, J. C. Penney Co.
Strategic Considerations to Store Location Analysis in Mexico

SESSION C6

ECONOMIC DEVELOPMENT PLANNING IV

Organizers and Chairs:
B. T. Cullen, University of New Mexico
L. M. Sommers, Michigan State University

A. Mehretu and **L. M. Sommers**, Michigan State University
Towards an Explanation of Micro-Marginality in Southeastern Michigan

W. J. Gribb, University of Wyoming
Land Consolidation as a Tool for Development on the Wind River Reservation

R. J. Czerniak, New Mexico State University
The North American Free Trade Agreement and its Impact on the Development of the Central Rio Grande Valley

B. T. Cullen and **R. L. Curtiss**, University of New Mexico
Location Decisions for Franchisees and Franchisors

SESSION D4

GIS II

Organizer and Chair:
J. Morrison, USGS
Urban Applications of Geographic Information Systems

R. DeAngelis, USGS/NMD
National Capital (Washington, D.C.) - PlanningProject

R. Birnknopf, USGS/GD
Economic and Statistical Analysis on the Risk of Different Earthquake Magnitudes on Structures

D. Icove, FBI
Serial Violent Crimes Project using GIS

7:15-8:45 p.m.

SESSION A3

RETAIL MODELS I

Organizer and Chair:
J. H. Haake, Fay's Incorporated

P. Densham, National Center for Geographic Information and Analysis, State University of New York at Buffalo
A Spatial Decision Support System for Optimizing Retail Bank Delivery Networks

T. Meyer, MarkeTech Systems
Optimal Network Location Strategies Using Site

J. Clarke, Manchester Metropolitan University, Manchester, England
Managing Saturation: Examples of the Use of Locational Strategy & Analysis from the U.K.

SESSION C7

COMPUTER MAPPING APPLICATIONS IN THE UTILITY INDUSTRY: BEYOND AM/FM

Organizers and Chairs:
J. W. Frazier, SUNY, Binghamton
L. Freeman, Applied Energy Group, Washington, D.C.

A. Davis, GeoDemographics, Limited
T.B.A.

L. Freeman, Applied Energy Group
T.B.A.

SESSION C8

ECONOMIC DEVELOPMENT AND PLANNING V

Organizer and Chair:
I. Lindsay, Ryerson Polytecnical Institute

J. A. McIntyre, Economic Development Corporation, Illinois
The Illinois & Michigan Canal as an Economic Development Tool

M. Green, University of Western Ontario
Canadian Institutional Development and Regional Development

SESSION E1

CAREERS IN GEOGRAPHY

Organizer and Chair:
R. Janson, The Janson Industries

PANEL DISCUSSION:

R. Janson, The Janson Industries
L. Carlson, Dayton Hudson Corporation
H. Stafford, University of Cincinnati
L. Sparks, University of Stirling, Scotland

FRIDAY, October 15, 1993

9:00-12:00p.m.

THEME SESSION III - NAFTA

Organizer and Chair:

D. Mock, Ryerson Polytecnical Institute

PANEL DISCUSSION:

D. Crane, Toronto Star
J. Felt, United States Consulate
A. Nieto, Mexican Consulate
J. Myers, Canadian Manufacturers Association

DISCUSSANT:

J. N. H. Britton, University of Toronto
8:00-9:30 a.m.

SESSION A4

RETAIL PLANNING II

Organizer and Chair:
B. Epstein, Kent State University

S. Sampson, Babson College
T.B.A.

D. P. Brennan, University of St. Thomas
Consumer Perception, A Small Town Retailers, Discounters, and Retail Communities

W. Dakan, University of Louisville
Site Analysis, Decision-Making and The Analytical Hierarchy Process.

P. S. Kopel, Kopel Research Inc.
T.B.A.

SESSION C9

ECONOMIC DEVELOPMENT PLANNING VI

Organizer and Chair:
N. Henry, Binghamton University, Binghamton, New York

J. A. McIntyre, Economic Development Corporation, Illinois
The Illinois and Michigan Canal as an Economic Development Tool

A. Clark, Department of Geography, East Carolina University
Spatial Changes in Banking Operations in North Carolina

D. Jardine, GeoDemographics, Limited
Trade Lead Matching and Small Business Development in Appalachia

SESSION B5

ENVIRONMENT POLICY III

Organizer and Chair:
D. Alvic, University of Tennessee
T.B.A.

10:00-11:30 a.m.

SESSION A5

RETAIL PLANNING III

Organizer and Chair:
D. Huntoon, Thompson and Associates

T.B.A.

SESSION C10

MEDICAL GEOGRAPHY

Organizer and Chair:
R. Gordon, Arizona State University

L. A. Thompson, Office of the Medical Director and **S. McGraw-Helms**, Arizona Department of Health Services
A Neighbourhood Outreach Program: Demographic Factors Affecting Site Selection and Adaptation

D. A. Hanson, Good Samaritan Regional Medical Center
Using Arizona Zip Codes to Target Areas of High Incidence of Baby's Exposed to Drugs in Utero

B. Hoefle, INTELLIMED and **R. J. Gordon**, Arizona State University
Using Public Hospital Discharge Data and GIS For Informed Medical Decision Making

G. Wilcox, University of Windsor
An Investigation Into the Location of the Freeport Chronic Care Hospital in Kitchener, Ontario

F. C. Innes, University of Windsor and **I. Cinolino**, Sandwich Community Health Centre
A Holistic Approach: From Health Needs Survey to Community Development: A Windsor, Ontario Survey

SESSION D5

CENSUS DATA BASES AND DEFINITIONS

Organizers and Chairs:
J. Sobel and **N. Torrieri**, U.S. Bureau of the Census

T. Haythornthwaite, Statistics Canada
Geographic Research for the 1996 Census

J. Sobel, U.S. Bureau of the Census
Using Locally-Maintained Digital Geographic Files to Update the TIGER Data Base

N. Torrieri, U.S. Bureau of the Census
The Population Density of Urbanized Areas

M. Ratcliffe, U.S. Bureau of the Census
Small Urban Agglomerations

SESSION B6

APPLIED RESEARCH IN THE NORTH

Organizer and Chair:
F. Duerden, Ryerson Polytecnical Institute

T.B.A.

12:00-1:30 p.m.

LUNCH

1:45-3:15 p.m.

THEME SESSION IV - ONTARIO WASTE MANAGEMENT ISSUES

Organizer and Chair:
R. Pushchak, Ryerson Polytecnical Institute

B. Massam, York University
Bill 143, Waste Management in the Toronto Region and Decision Support Systems

D. Lang, The Toronto Harbour Commission
The Toronto Harbour Comissioners' SoilRecycling Demonstration Project

B. Bardswick, Ontario Ministry of the Environment
Regional Planning Strategy for Biomedical Waste Disposal in Ontario

1:45-3:15 p.m.

SESSION A6

RETAIL MODELS II

Organizer and Chair:
A. Lea, Compusearch Micromarketing Data & Systems

A. Newell, Spectrum Decision Sciences
Network Optimization Models

K. Lee, Spectrum Decision Sciences
Geographic Market Potentials Using a Geodemographic Market Segmentation System

D. Blanchett, Compusearch Micromarketing Data & Systems
Marketers Use of GIS Technology

SESSION B7

WATER RESOURCES III

Organizer and Chair:
G. Tobin, University of Minnesota

T.B.A.

SESSION C11

URBAN AND REGIONAL PLANNING ISSUES

Organizer and Chair:
M. Kosny, Ryerson Polytecnical Institute

M. Kosny, Ryerson Polytecnical Institute
Development: New Challenges and NewResponses from the Canadian Planning Front

L. Altrows, Ryerson Polytecnical Institute
Strategic Planning as an Approach to Regional and Community Development: Four Canadian Cases

J. Springer, Ryerson Polytecnical Institute
The Impacts of Growth Management and Affordable Housing in the Ontario Context

D. Amborski, Ryerson Polytecnical Institute
Controlling Development in the Greater Toronto Area

SESSION D6

GIS III

Organizer and Chair:
H. Veregin, Kent State University

L. Betts, Kent State University
Equity Flow Problems in the Context of Regional Development

L. Clark, Kent State University
Delineation of the Ohio and Erie Canal Corridor using GIS

J. Moeller, Kent State University
Dynamic Modeling of Wetlands using GIS and Remote Sensing

R. Pandya, H. Veregin and R. Roethlisberger, Kent State Univesity
An Interactive System for Visualization and Exploration of Spatial Data Quality

H. Veregin and F. Sultana, Kent State University
Resolution-Dependent Effects on Classification Accuracy in Remote Sensing

3:30-5:00 p.m.

SESSION D7

EMERGENCY MAPPING

Organizer and Chair:
U. J. Dymon, Kent State University

U. J. Dymon, Kent State University
Communication of Spatial Data During Hurricane Andrew

N. L. Winter, Clarke University
A Map Taxonomy for Emergencies

M. Monmonier, Syracuse University
Graphic Narratives for Emergency Mapping

SESSION A7

RETAIL PLANNING IV

Organizer and Chair:
D. Huff, University of Texas

J. D. Lord, University of North Carolina at Charlotte
Power Shopping Centers and Power Politics: Retail Development in Charlotte, NC

G. Featherstone, Dylex Ltd.
The Role of Business Strategies in the Development and Implementation of Retail Location Models

SESSION E2

GEOGRAPHIC EDUCATION IN ONTARIO

Organizer and Chair:
G. Goodfellow, Toronto Board of Education

PANEL DISCUSSION:

M. Oliver, Napanee Secondary School
S. Van Geest, Applewood Heights Secondary School

6:00-10:00p.m.

BANQUET AND RECEPTION

GUEST SPEAKER:

Hon. **R. Grier**, Minister of Health, Province of Ontario

SATURDAY, October 16, 1993

9:00-12:00 p.m.

SESSION F1

MILITARY GEOGRAPHY

Organizer and Chair:
Col. R. C. Ham, United States Military Academy, West Point

Capt. M. W. Corson, United States Military Academy, West Point
Strategic Mobility in the Gulf War

Maj. C. W. Swanson, On-Site Inspection Agency, Washington, D.C.
The Devolution of the Soviet Union and Its Impact on Nuclear Arms Control Treaty Implementation

Col. R. C. Ham, United States Military Academy, West Point.
U.S. Security Interests in Southeast Asia

K. A. Schroeder, University of Minnesota - Duluth, Minnesota
Operational Art in the Bosnian War

SESSION C12

INTERNATIONAL ISSUES AND ECONOMIC DEVELOPMENT

Chair:
S. Swales, Ryerson Polytecnical Institute

K. Mueller, Kent State University
Tindorama: European Agriculture Settlement Scheme, Coastal North-East Brazil

M. Nyamwange, East Stroudsburg University
Towards Environmental Rehabilitation in Africa

SESSION C13

THEMES IN APPLIED GEOGRAPHY

Organizer and Chair:
M. E. Harvey, Kent State University

C. L. Cameron, Kent State University
The Adoption of Geography Courses in 2-Year Colleges: A Systems Approach

K. Gregory, Kent State University
Identifying Agricultural Non-Point Source Pollution Hotspots: The Aggregation Problem

S. Tripathi, Kent State University
Decision Making Behavior About Sexually Transmitted Diseases by Rural Women in India: Implications for Planning

R. Janson, The Janson Industries
Ohio's Defense Conversion Policy: A Prescription for Future Economic Growth

M. E. Harvey and **E. Tettey-Fio**, Kent State University
Dynamic Analysis of Interregional Migration: Implications for Regional and National Planning

9:30-12:00 p.m.

THEME SESSION V - TORONTO WATERFRONT REGENERATION

Organizer:
S. Barrett, Waterfront Regeneration Trust, Toronto
T.B.A.

12:00-3:00 p.m.

FIELD TRIPS
- Toronto Island
- Toronto Underground Mall
- Waterfront
- Cultural Toronto
- Historic Toronto

ACKNOWLEDGEMENTS

We gratefully acknowledge the continuing financial support for the Applied Geography Conferences from Binghamton University, Kent State University, and the University of North Texas. We express our appreciation to Ryerson Polytechnical Institute and its School of Applied Geography for their excellent services and support as host of the 1993 Applied Geography Conference.

Special thanks are due to our adminisrative agency for 1992-93, Van W. Emmons, Sr. Population and Marketing Analysis Center (PMAC), and their staff. They also prepared this manuscript for printing.

Finally, we express sincere thanks to our Steering Committee for their many services in 1992-93 and to all Conference organizers, chairs, field trip leaders, speakers, student workers and others who contributed their time freely to make this sixteenth Applied Geography Conference a success.

PINDORAMA: EUROPEAN AGRICULTURAL SETTLEMENT SCHEME, COASTAL NORTHEAST BRAZIL

Keith D. Muller
Department of Geography
Kent State University-Trumbull Campus

1. INTRODUCTION

Poverty in Northeast Brazil has been widespread since European settlement began in the 1500s. Many of today's 45 million people chose, especially in years of drought, to migrate to the South in search of economic opportunity. The physiological population density prompts this migration. Traditionally, the fertile valleys of the coastal region have been dominated by large sugar cane operations. However, one portion has generally been ignored: the *tabuleiros,* the flat tablelands, 50 to 200 meters above sea level, 20 to 40 kilometers wide, 50 to 200 meters above sea level, 20 to 40 kilometers wide, extending 750 kilometers along the Atlantic coast. One attempt to utilize the *tabuleiros* is the Pindorama settlement scheme in the state of Alagoas, 250 kilometers south of Recife. The focus of this study is the development of this experiment, and how it may provide a model for future growth in the Northeast, as well as throughout Brazil and in other developing nations.

2. THE FORMATIVE YEARS

Since the 1500s, the coast of the Northeast has been dominated by sugar cane cultivation, mostly confined to fertile river valleys, on large *latifundia* holdings. Because of soil conditions, sugar cane was not cultivated in the less fertile *tabuleiros*, which had been only sparsely utilized for cattle grazing, lumbering and subsistence agriculture. (2,3) Thus, the densely populated Northeast had, with the *tabuleiros,* a potential location for the establishment of settlement schemes. With this in mind, a Brazilian company, *Companhia Progresso Rural* (Rural Progress Company), established the Pindorama project in 1954 with the purchase of 34,133 hectares of land on the *tabuleiros*. By 1958, the company discontinued operations due to financial problems despite the construction of 200 houses, 150 kilometers of roads, urban headquarters and the selection of the firs 190 settlers. In 1959, the tasks of settlement and the marketing of the agricultural production were assumed by *Cooperativa de Colonizacao Agropecuaria e Industrial Pindorama Limitada* (Pindorama Agricultural Colonization and Industrial Cooperative, Limited), which expanded the consumers' cooperative that was founded in 1956. By 1961, 420 lots were occupied and the Company procured a Brazilian federal loan for US $98,000 to build a juice factory. This factory processed passion fruit (*passi flora edulis*), which gave the project a new emphasis to market a value-added crop (both grown and processed locally) to the project. (1,5).

By the early 1960s, German and Swiss officials were financially and administratively involved. Thus, relatively large amounts of foreign capital were invested in this experiment. Despite some precarious years from 1962 to 1974, international technical and financial aid continued to arrive from the United States (including Pease Corps Volunteers), West Germany, The Netherlands and Switzerland. Over 100 foreigners a year worked in Pindorama from 1961 to 1970. By 1970, the project had received an estimated US$5 million in international aid, or about US$12,000 per farm, which is certainly a high figure for any place in the world. Moreover, the original goal of settling 1,500 individual farm lots was only half met at this date, despite such large sums of aid. (1,5) It was not until 1976 that the first Brazilian governmental agency, Superintendencia de Desenvolvimento do Nordeste, (SUDENE), Superintendency of Development of the Northeast was founded.

3. SETTLEMENT, SITE AND SITUATION

About one-half of the project lies on the *tabuleiros* at elevations varying from 50 to 150 meters. The remaining area consists of dissected *tabuleiro,* which has been carved out by the Piaui River and its tributaries: the V-shaped valley reaches 5 meters above sea level at a point ten kilometers from the Atlantic Ocean.

The project's principal road is unpaved and connects with the asphalt state highway located only three kilometers from the project's eastern edge -- or six kilometers from the urban center of Pindorama. The lack of an asphalt road has been a major detriment to the project throughout its 40 years of existence. Pindorama is especially isolated during 45 to 60 days of the rainy season, usually between April through August, when only tractors and four wheel drive vehicles can pass over the road. Continuous rainfall may extend for up to ten days; and even during the dry periods the road is bumpy and dusty.

To improve the chances of state funds for the eventual construction of an asphalt road to Pindorama, the colony has campaigned for the status of *municipio. Municipios* are comparable to counties in the United States except that new *municipios* are often formed once an urban center of "significant" population has developed and a sound tax base is shown. However, politicians and landowners of large properties surrounding the colony, often one and the same, are threatened by any type of success for Pindorama. Thus, all attempts to construct an asphalt road and to become a *municipio* have failed. Moreover, the Pindorama project is located in portions of four *municipios* making it of secondary concern to each of these. Even though the urban center of Pindorama lies in the *municipio* of Cururipe, and most of the taxes from the project are paid there, officials are not sympathetic to the interests of Pindorama.

4. SETTLEMENT, FORM AND STRUCTURE -- THE LONG-LOT SYSTEM OF LAND DIVISION

In the Northeast, Pindorama is unique in its system of land division. Although long-lots (lengths two or more times greater than widths and rectangular in form) are meagerly scattered throughout the Northeast, only Pindorama has a well-planned long-lot land division system.

Project maps show that two basic types of surveyed long-lots dominate Pindorama. Type I, located east of the city of Pindorama, is composed of non-topographically adjusted long lots surveyed in rectangular blocks that were established during the early years of the project -- in the 1950s and early 1960s. This settlement type covers about 25 percent of the project and is confined mostly to the flat tabuleiros. Most lots consist of 20 to 30 hectares of land and range from the typical rectangular form (300 x 1,000 meters) to a few infrequent squares (600 x 500); or width to length ratios of 1:3.3 to 1:1 (a square).

This area of Type I lots also has three planned group settlements, or *agro-vilas,* established to minimize isolation f each farm family. However, those settlers who reside in the *vilas* must commute as much as two kilometers to tend crops and animals. This is a clear disadvantage of the system. The three agro-vilas range from 50 to 150 families for a total of about 300 urban lots. In addition, from 50 to 60 families, or 20 percent of Type I lot owners have chosen to live on their lots. Some settlers formerly lived in *agro-vilas,* but sold their urban house to relocate on their agricultural lots, which permits improved crop and animal care.

The Type II lot is a topographically-adjusted long lot, i.e., extending from the inerfluves of the dissected *tabuleiros* to the streams below. Their width to length ratios of 1:6 to 1:10, or 50 x 300 meters to 100 x 1,000, are greater and vary more than Type I lots which more resemble squares in their dimensions. The Type II lots were gradually surveyed and settled, from the late 1960s to the present, and comprise about 65 percent of the project's area.

Type II lots offer a relatively equal distribution of soils, micro-climates and various degrees of slope each holding. Admittedly, settlers are somewhat more isolated compared with Type I settlers and their *agrovilas*. However, since each family resides on its individual lot, Type II settlers can devote closer attention to their crops and animals. Moreover, the distance between neighbors is minimized by the line settlement pattern that results. Farmsteads tend to be strung along the roads at the end of the adjoining lots, usually, on the interfluves, but occasionally at the valley bottoms. These topographically-adjusted long-lots are similar to rural settlements in Europe such as the Waldhofendorf, forest long-lot village, as well as those found throughout southern Brazil. (4) From the total of over 1,300 farms on Pindorama, Type II has about 900 farms and Type I about 400, or 70 and 30 percent, respectively.

5. SETTLEMENT SIZE

Type II lots, which were established in the later stages of development, normally are smaller by one-half to one-third the size of Type I lots surveyed in the initial stages of the project, or approximately 10 to 20 hectares versus 20 to 30, respectively. Table 1 shows holding sizes for the entire project. Few small holdings exist and the largest holding is only 105 hectares. 32 holdings, or 3.2 percent, have less than ten hectares, while 90.6 percent are from 10 to less than 30 hectares and only 6.2 percent are 30 or more hectares in area. According to the cooperative statistics, Pindorama has 1,386 rural lots comprising 32,309 hectares, or an average area of 23.3 hectares per lot. There are 472 urban lots in the project's urban center of Pindorama.

Table 1

CLASSES OF HOLDING SIZE, PINDORAMA

Size (Hectares)	# of Holdings*	Percent of Total
<5	32	2.4
5<10	11	0.8
10<15	339	25.4
15<20	364	27.2
20<25	177	13.2
25<30	332	24.8
30<35	63	4.7
35<40	7	0.5
>40	13	1.0
Totals:	1,338	100.0

Cooperative records.
*48 lots are not accounted, 29 were not classified and four had no owner listed.

Coruripe, the municipio in which about one-half of the project lies, shows the extent that large properties dominate outside of Pindorama; Table 2. For example, 72.9 percent of the area is in properties of more than 100 hectares, and 33.6 percent more than 1,000. Moreover, it is common in coastal Northeast for single individuals to control large tracts of land than official statistics fail to reveal. This practice is disguised by using the names of relatives on land titles. As a result, individuals and their families may control tens of thousands of hectares. Thus, the success of a project such as Pindorama is a threat to their social, cultural and economic control. Conversely, this Pindorama experiment shows that through unity and cooperative effort small farmers may be able to escape the cycle of poverty common to the Northeast and gain a respectable standard of living.

Table 2

CLASSES OF HOLDINGS, MUNICIPIO OF CURURIPE, ALAGOAS

Size (Hectares)	# of Holdings	% of Holdings	Total Area Holdings	% of Area
<10	1,662	65.2	2,716	3.3
10<100	778	30.5	19,765	23.8
100<1,000	99	3.8	32,630	39.3
1,000<10,000	12	0.5	27,971	33.6
Totals	2,551	100.0	83,082	100.0

Censo Agropecuario do Brasil - Alagoas, 1985.

Table 3

**HOLDING CLASSES BY SIZE AND AREA,
CORURIPE, NORTHEAST BRAZIL**

Class Size (Hectares)	# of Holdings*	% of Holdings	Total Area (Hectares)	% of Area
<10	1,662	65.2	2,716	3.3
10<20	334	13.1	4,847	5.8
20<50	337	13.2	10,466	12.6
50<100	67	2.6	4,471	5.4
100<200	41	1.6	5,844	7.0
200<500	36	1.4	11,666	14.0
500<1000	22	0.9	15,120	18.2
1000<2000	8	0.3	11,979	14.4
2000<5000	3	0.1	7,979	9.6
5000<10000	1	<0.1	8,738	10.5
Totals:	**2,554**	**100.0**	**83,082**	**100.0**

*Three holdings are undeclared.
Sinopse Preliminar do Censo Agropecuario, Alagoas, 1985

6. THE COOPERATIVE

From its inception, the main objectives of the cooperative were to complete settlement of the project and to process and market agricultural production. The first attempt to process products for the members, and thus add value, occurred in the early 1960s with the installation of the passion fruit factory that received all of the fruit produced in the settlement for concentrated juice and preserves. The preserves were later abandoned, but passion fruit juice production continues today. In addition, about 25 percent of the pineapple grown in the project is also processed; the remainder is sold as fresh fruit at local markets. The juices are marketed exclusively in Rio de Janeiro and Sao Paulo. Although coconuts from the project have been marketed throughout its history, only in 1993 were funds available for a coconut processing factory.

Until the early 1970s, the cooperative's activities centered on the passion fruit juice production and the local marketing of coconuts. However, with the federal gasohol program in the 1970s for sugar cane, the cooperative built a distillery (1981) capable of producing 240,000 liters per day. As long as sugar cane commanded high prices, passion fruit became secondary. Today they are equally valuable to the project. Diversification is important to the cooperative so as not to be affected by fluctuating prices and possible crop failure associated with a monoculture.

The cooperative not only guarantees its members the purchase of most of their products and a share in the profits, but it is also active in providing services such as schools, medical facilities and agricultural equipment. The cooperative maintains exclusive ownership to the land. Its members only have the right of possession to their lots. Certainly, this is beyond the normal role of cooperatives in Brazil. Possession is transferable by inheritance and the right to sell to anyone, including non-family members, is allowed with the approval of the cooperative. However, possessors must live in the project and personally work the land. This prevents absenteeism and insures the cooperative control of fulfilling the original objectives of the cooperative. A few initial settlers have outright ownership since they had completed payments in full before the cooperative assumed administration of the project in 1959.

7. SETTLEMENT FUNCTION -- ECONOMIC LIFE

Although the *tabuleiros* were considered by the sugar cane growers as of no use, small farmers had sporadically raised staple crops: manioc, maize and beans. The fact that sugar cane growers considered the *tabuleiros* as of no use, certainly, meant that inexpensive land, located close to the Atlantic coast, was available to a project such as Pindorama in the 1950s and 1960s. However, since the early 1970s, large sugar cane operations, responding to the subsidized federal *Pro-alcool* program, began t convert sugar cane to alcohol for gasohol. This began large-scale sugar cane production on the *tabuleiros* -- especially where mechanized, modern agricultural techniques with chemical fertilizers were applied to correct soil deficiencies.

The local consequence of the *Pro-alcool* program was the expansion of sugar cane production, almost exclusively to the *tabuleiros*. For example, in Coruripe production increased 800 percent, from about 6,000 hectares in 1970 to over 55,000 hectares in 1985; Table 4. In the same period, the state of Alagoas increased production 200 percent from 166,194 to 496,709 hectares. The increase in Coruripe to over 30,000 hectares between 1970 and 1980 is mostly within the non-Pindorama portions of the *municipio*. The 1981 opening of the Pindorama distillery accounts for a large share of the 22,000 hectare increase between 1980 and 1985. Project data from 1990 show that members planted a total of 12,000 hectares of cane. Table 5 shows that in 1992 over 400,000 tons of cane produced more than 31 million liters of alcohol.

Table 4

SUGAR CANE AREA, CORURIPE
NORTHEAST BRAZIL

Year	Area in Hectares
1970	6,232
1975	18,762
1980	33,352
1985	55,709

Censo Agropecuario do Brasil - Alagoas, 1970, 1975, 1980, 1985

Although many settlers shifted to sugar cane at the expense of other crops in the 1980s, the cooperative realized the risk of over-dependency on a crop subject to dramatic price fluctuations and possible crop failures. Thus, one objective of the cooperative is to maintain diversity in production. For example, a new coconut factory was completed in March, 1993 which can process up to 8,000 coconuts a day. Also in 1993, funds were received to construct a dairy creamery to process milk. Thus, the gasohol distillery, the passion fruit, pineapple, and coconut factories and the creamery are all attempts at diversification by processing and marketing products that add value and provide non-agricultural employment.

Table 5

SUGAR CANE PRODUCTION, PROCESSED CANE, ALCOHOL PRODUCTION PINDORAMA, 1990 TO 1992

	Processed Cane (Tons)	Alcohol Production (Liters)
1990	486,247	35,787,419
1991	396,253	31,314,461
1992	403,610	31,896,247

Cooperative Records

In addition to crops mentioned above for processing and marketing by the cooperative; manioc, beans, rice, cashews and mangoes are grown for subsistence or are marketed locally.

8. Conclusion

Northeast Brazil has been troubled throughout its existence by poverty and drought. Since European settlers first arrived in the 1500s, the population has lived mostly on the coast and much of the agriculture was confined to the fertile valleys in the area. However, one of their most valuable land resources, the *tabuleiros,* has been practically unused for the past four centuries. Utilization of this area may allow the inhabitants to break the cycle of poverty and despair and become contributing members of a better society.

One attempt to utilize the *tabuleiros* is the Pindorama settlement project. The system of land division closely resembles the successful European and southern Brazilian long-lot system. This project demonstrates that it is possible to develop the sparsely populated *tabuleiros* and improve the lives of the settlers through the formation of a cooperative. The project began in 1954, and within 10 years was the target of a massive amount of foreign aid. The move away from subsistence farming to a more profit yielding system was made possible by the construction of processing facilities near the settlement. Now the cooperative runs a juice factory, a creamery and a sugarcane processing factory that produces alcohol for the nation's gasohol program. In 1993, separate funds were allocated for the construction of a coconut factory and for a creamery to process milk.

Sugar cane production has had an enormous effect on the use of the *tabuleiros* of Northeast Brazil. With the expansion of production to meet government goals for gasohol production, not only large operations, but also settlement.

Organizations such as the Pindorama with small farmers can benefit through cooperative efforts. This promotes the betterment of their lives through commercial rather than only subsistence farming.

There are some difficulties facing the settlement. A paved road to link the settlement with the main highway, a mere 6 kilometers from the urban center of Pindorama, has not been constructed, despite repeated attempts to have it approved. Efforts to become a *municipio,* the Brazilian equivalent to a U.S. county, has been stymied by local politicians and powerful landowners. These factors have hindered the success of the program, but have not yet led to its demise.

Poor farmers in the developing world, as a general rule, barely survive from agriculture. However, with the application of cooperative effort, real improvement in production and subsequently, standard of living may be possible. Successful cooperatives provide valuable examples to other nations. The Pindorama settlement is a model for the future of Northeast Brazil and developing countries throughout the world. Through cooperative action in Pindorama, settlers have raised their standard of living and have moved beyond the subsistence agriculture that grips poor farmers in Northeast Brazil.

9. REFERENCES

1. *Cooperativa de Colonizacao Agropecuaria e Industrial Pindorama Limitada* (Pindorama Agricultural Colonization and Industrial Cooperative, Limited), no date. "Breve Historico da Empresa".

2. Correa, R., 1963. "A Colonia de Pindorama: uma modificacao na paisagem agraria dos Tabuleiros Alagoano, Revista Brasileira de Geografia. 479-484.

3. Heredia, A. de, 1988. Formas de Dominacao e Espaco Social: A Modernizacao da Agroindustria Canavieira em Alagoas. Sao Paulo: Marco Zero.

4. Muller, K. D., in press. O Impacto de Mecanizacao no Sul do Brasil: Caso do Oeste do Parana. Rio de Janeiro, Brazilian Institute of Geography and Statistics.

5. Schacht, S. 1980. "Agrarkolonisation in der *Zona da Mata* Nordostbrasiliens am Beispiel der Kolonie Pindorama", 54-76. In English, 1983? "Agricultural Colonization of the *Zona da Mata* of Northeast Brazil -- The Example of Pindorama", Applied Geography and Development. 71-90.

BENEFITS AND PROBLEMS IN USING A SIDE-CHANNEL RESERVOIR CONFIGURATION, CHARLESTON, IL

Vincent P. Gutowski
Department of Geology/Geography
Eastern Illinois University
Charleston, IL 61920

1. INTRODUCTION

As cities expand, so do their needs for a reliable source of potable water. Each geographical setting is unique and the ability of the local region to meet water needs often is taxed as population expansion takes place. Innovative means of water withdrawal and/or storage are sought that will meet present and future water needs. Charleston, Illinois is an area which has undergone change in the last half-century that has prompted both traditional and more recently, non-traditional methods for meeting water demand. Each method has, at first, seemingly improved the quantity and quality of the water supply, but eventually new sets of problems have emerged with time.

Charleston is located in east-central Illinois (Figure 1), an area characterized by thick glacial deposits and spotty groundwater resources unsuitable for municipal withdrawals. Most towns rely on surface water impoundments for adequate water supplies. Charleston's first impoundment was a dam across the Embarras River in 1895. The dam, two miles east of town, kept the river at depth of 6 feet throughout the year. This proved adequate for nearly a half century until population pressures required a larger impoundment.

Riverview Dam was completed in 1947, creating Lake Charleston, a wide, shallow lake with a total capacity of approximately 744 million gallons (Figure 2-A). Because the Embarras River drains just over 811 square miles of predominately agricultural land, sedimentation occurred at a rapid rate. Research by the Illinois State Water Survey indicated that by 1960 the capacity was reduced to 455 million gallons and in 1974 the capacity was further reduced to 282 million gallons. (1) Thus, in 27 years the lake has lost 62 percent of its original capacity.

To remedy the loss of storage capacity due to siltation, a new reservoir configuration was designed and built by 1980 and eventually filled to capacity (1,200 million gallons) by 1982. City officials expected that Charleston would have a large quantity of high-quality water, sufficient to meet the needs for future industrial, domestic and municipal growth.

2. RESERVOIR SETTING

The new reservoir design was a "side-channel reservoir" type, requiring a portion of the floodplain be isolated from the main river channel by building a mile-long dike with a crest approximately 15 feet higher than the river surface elevation (Figure 2-B). The Embarras River would flow to the east side of the dike and remain impounded by the dam and spillway constructed in 1947. The new reservoir would be kept full by runoff from the smaller, surrounding watershed and by selective pumping from the impounded river. By pumping into the reservoir only when river water was relatively free from suspended load and by allowing sediment-laden flow to bypass the reservoir, it was assumed that the new reservoir would not suffer from siltation as it had in the past. This method of reducing sediment inflow to the reservoir is one of the highly recommended methods of water storage conservation. (2)

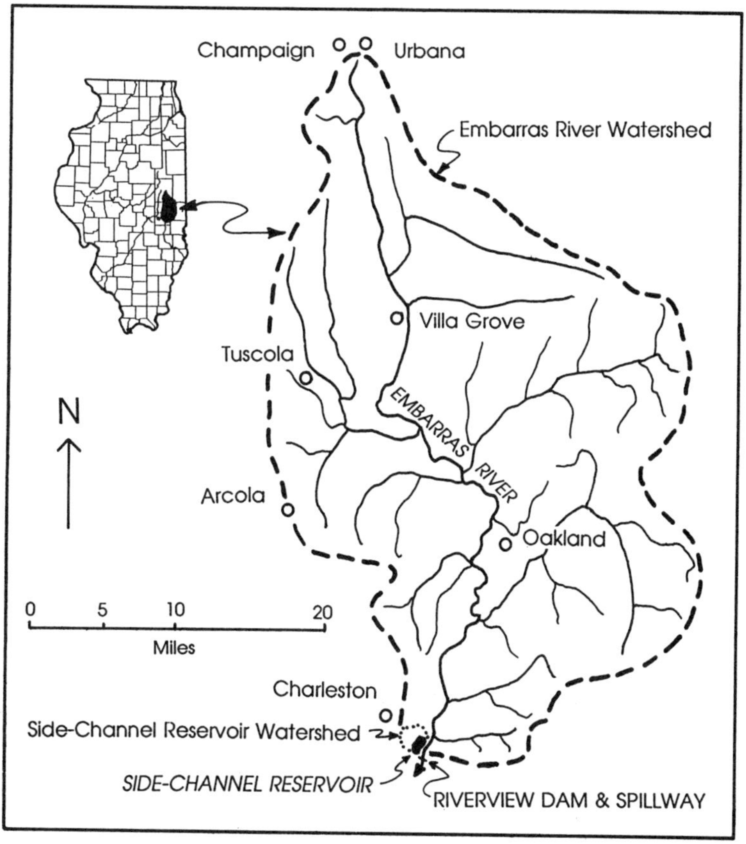

Figure 1. Location map of Charleston, IL, showing the Embarras River watershed upstream from Riverview Dam and the watershed that drains directly the Side-Channel Reservoir.

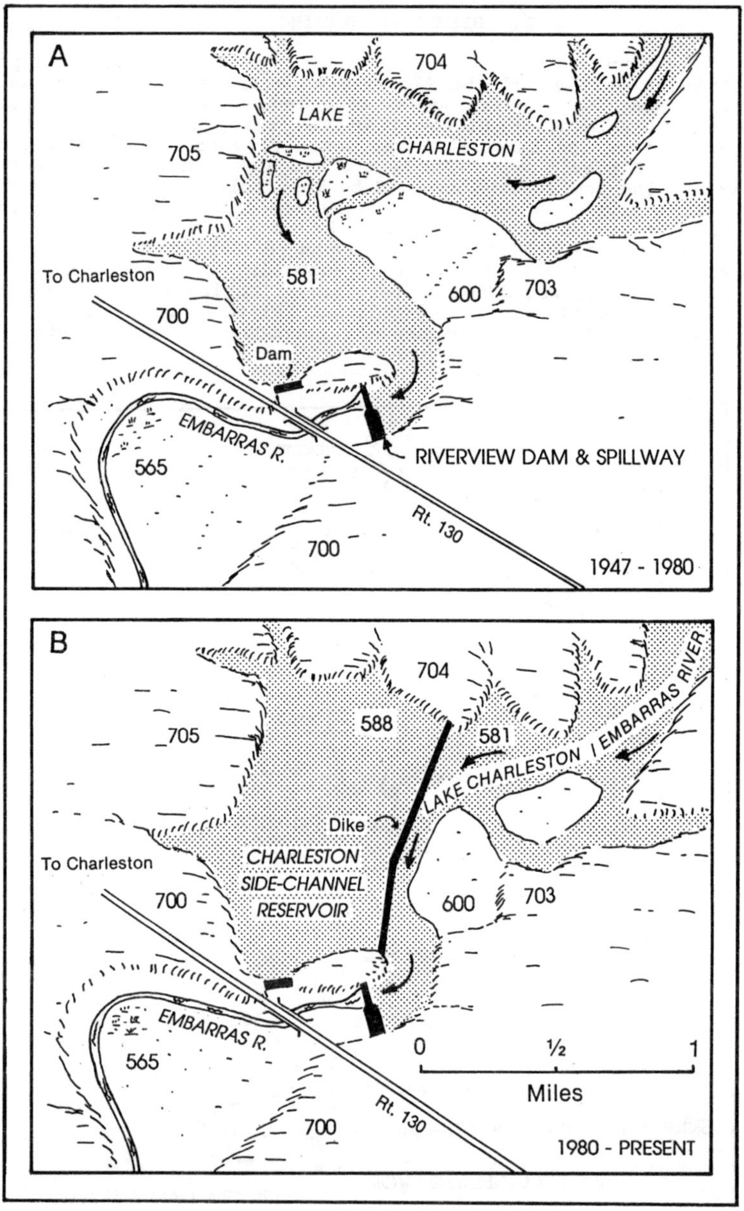

Figure 2. Configuration of the Charleston, IL, municipal water impoundment: A) Lake Charleston, 1947-1980; and, B) Charleston Side-Channel Reservoir, 1980 to present time. Arrows show the direction of water flow from the Embarras River watershed.

Construction of the side-channel reservoir had the unique effect of creating a reservoir with two watersheds (Figure 1). By isolating the reservoir from the Embarras River, the total area that drains directly into the reservoir was reduced from the original 811 square miles to just 1.8 square miles, this figure includes the 0.51 square mile reservoir surface (Figure 3). With a dual watershed configuration, the city could maintain greater control over the quality of water captured for storage. Approximately 49% of the water introduced into the side-channel reservoir comes from the immediate watershed, via direct rainfall into the reservoir or from surface and subsurface flow. Pumping from the impounded Embarras River accounts for 51% of the water input. If the city could regulate the timing of the pumping from the river and could supervise the land use in the reservoir watershed, it would provide enhanced control over the quality of its municipal raw water supply.

CHARLESTON SIDE-CHANNEL RESERVOIR WATERSHED

LAND USE MAP

LEGEND	MI^2	ACRES	% OF TOTAL
A=AGRICULTURAL	.1450	92.8	8.2
C=COMMERCIAL	.0225	14.4	1.3
G=GRASSLAND	.1050	67.2	5.9
I=INDUSTRIAL	.0100	6.4	.6
F=FOREST	.7775	497.6	43.8
M=MUNICIPAL	.0150	9.6	.8
RE=RESIDENTIAL ESTATE	.0925	59.2	5.3
RI=RESIDENTIAL	.0900	57.6	5.1
W=WATER	.5125	328.0	29.0
TOTAL	1.770	1132.8	100.0

ALL BOUNDARIES DRAWN FROM USGS TOPOGRAPHIC MAPS AND AIR PHOTOS. V.P.G 4/3/88

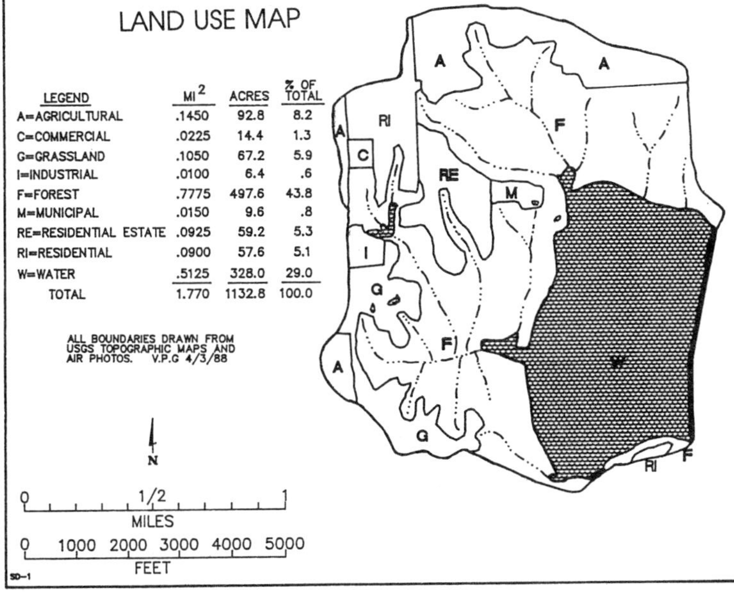

Figure 3. Map showing the land use in the watershed that drains directly into the Side-Channel Reservoir.

3. BENEFITS

Changing the configuration of Charleston's raw water supply from a typical dam and spillway river impoundment to a side-channel reservoir has produced some benefits and some drawbacks. Benefits include a larger volume facility for raw water storage. Currently there is approximately an 18 month supply of water stored in the reservoir, nearly a five-fold increase in water reserves. The increase in storage capacity should minimize problems during drought episodes.

There has been a dramatic reduction in sediment input values. Currently, annual sediment input is approximately 1/15th of pre-1980 values. However, although fluvial sedimentation has decreased, the amount contributed from shoreline erosion has increased.

A dual-watershed configuration allows water to be pumped from the river to the reservoir only when water conditions are optimal. Optimum conditions include river water that is relatively free from sediment, algae and harmful chemicals. In 1989, Recontec Corporation built a heavy metals reclamation facility near Oakland, IL, 15 miles north of Lake Charleston. Any runoff from the facility would enter the Embarras River and potentially be pumped into the side-channel reservoir. In 1990, the City of Charleston installed continuous monitors on the river just downstream from the Recontec facility and at the intake of the pumphouse that transfers water from the river to the reservoir; another monitor was installed on a bridge half-way between Oakland and Charleston. The monitors will trigger an alarm at the Charleston Water Treatment Plant allowing city personnel to disable the pumps if there is a release of harmful chemicals into the Embarras River.

Vigilance over activities in the side-channel reservoir watershed allows further control over the quality of water entering the reservoir. In 1993 the City of Charleston passed the Charleston Side-Channel Reservoir Watershed Land Use Ordinance. The ordinance guides, regulates and controls activities that may contribute to the introduction of chemicals or excessive sedimentation into the reservoir. Public response to the ordinance has been quite favorable.

4. PROBLEMS

Although the plan for a trouble-free water supply, that would serve Charleston for future generations, seemed in place and working, several problems became evident over the next few years. Many of these problems were the result of the side-channel reservoir being a nearly closed hydrologic system. Although the reservoir has two water inputs and an output to the water plant, the residence time for water in the reservoir is nearly two years, a considerable difference from the previous system which was an impounded river with continuous throughflow.

4.1 PHYSICAL PROBLEMS

4.1.1 Stream Sedimentation

Sedimentation, one of the original problems in Lake Charleston, is still a point of concern with Charleston Side-Channel Reservoir. Although it is not as great a problem as when the full flow and sediment load of the Embarras River was entering Lake Charleston, there is sediment being introduced into the reservoir from the immediate watershed. Small tributaries entering the reservoir carry sediment during stormflow, with the amount of sediment being related to land use and rate of flow. Sediment input and stream discharge of several tributaries entering the side-channel reservoir were monitored over an 14 month period from February, 1989, to April, 1990; these tributaries drain approximately 50% of the reservoir watershed. During this time period it was determined that the tributaries ran relatively clean for most of the year. Only during storm events did the streams carry significant amounts of sediment. Even after intense storm events the streams would carry sediment for no more than 24 hours, after that the waters ran very clean. Streams draining agricultural areas carried the highest average amounts of sediment (0.285 gm/L), while areas where the land use is residential or forest/grassland carried less than half that amount (0.12 gm/L). Additionally, the streams are intermittent, being dry, except during storm flow, for approximately six months of the year. It is estimated that the annual sediment input to the reservoir via stream transport is approximately 2,064 tons per year. (3)

4.1.2 Shoreline Erosion

Another source of sedimentation into the reservoir is from shoreline erosion. The bluffs that make up the north, west and south shores of the reservoir are comprised of glacial materials. As the elevation of the water surface was increased by approximately 9 feet after construction of the reservoir, the waves began to work on the former valley side slopes. Slope angles are commonly at the limiting angle for soil-mantled slopes (27 to 32 degrees). As wave activity removes the topsoil and begins to work on the underlying glacial till, high angle (60 to 90 degree) unvegetated cliffs are formed. In some areas the cliffs are over 50 feet in height. These cliffs began to form immediately after construction of the reservoir and continue to develop today.

A vigil network was set up in 1988 to monitor the rate of cliff retreat. Between 1988 and 1993 the average rate of cliff retreat was approximately one foot per year. A related study done in 1992-1993 confirmed that the cliffs were retreating at this rate and actual retreat was not a slow, progressive retreat, but a seasonal activity confined to a few months or weeks during winter and spring. During summer and fall, cliff retreat is minimal; during winter months the exposed cliff surface is crumbled by frost activity to a depth of several inches. Winter rains and mass movement strip the loosened surface materials and deliver it to the base of the cliffs. This process can be repeated several times during winter months. Spring rains further strip any remaining material from the cliffs. Spring is the time of the year when the reservoir surface is at its highest point and storm-driven waves remove any loose material from the base of the cliffs. Materials eroded from the cliffs are carried into the reservoir, being deposited in a narrow band just offshore and the stage is set for another cycle of cliff retreat. A conservative estimate of sediment delivery to the reservoir due to shoreline erosion is approximately 1,000 tons per year.

4.1.3 Sediment Pumped From the Embarras River

Although water is pumped from the impounded Embarras River into the side-channel reservoir only when the river is carrying a relatively low suspended load, some sediment is pumped into the reservoir. Estimates of sediment loading due to pumping from the river to the reservoir are between 64 and 300 tons per year. Currently, total sediment input into the side-channel reservoir is an estimated 3,364 tons per year, a considerable improvement over the average of 52,500 tons per year into Lake Charleston during the period from 1947 to 1974. (1)

4.2 BIOLOGICAL PROBLEMS

Physical problems of the side-channel reservoir are primarily concerned with problems of water quantity. Biological problems affect water quality and often are tied directly to some of the physical problems.

4.2.1 Nutrient Loading

An abundance of nutrients in the side-channel reservoir is central to several biological problems. Phosphorus, and, to a lesser extent, nitrogen loading comes from both internal and external sources. Internal sources of phosphorus are sediments on the floor of the reservoir and the cycling of phosphorus by aquatic organisms. Internal phosphorus loading amounts to approximately 1,464 pounds per year. External sources of phosphorous loading (pounds/year) include: shoreline erosion (1,404), pumping into the reservoir from the Embarras River (313), and from tributaries to the side-channel reservoir (115). Phosphorus in the reservoir water creates conditions favorable for the growth of blue-green algae. Their growth, eventual death and decomposition contribute to several water quality problems.

4.2.2 Algal Blooms

When algae populations exceed 600 cells/mL it is considered to be in a bloom condition. Analysis of water in the side-channel reservoir during late summer periods showed there were more than 20,000 cells/mL. Algae blooms are considered a nuisance for several reasons. They produce an aesthetically displeasing scum on the water body and have an unpleasant smell. Excessive algae can create a problem for fish habitat by reducing the transparency of the water, creating a problem for macrophyte vegetation growth. Also, their decomposition depletes the dissolved oxygen values in bottom waters further disrupting the benthic fauna. Algae blooms, especially the blue-green type contribute significantly to taste and odor problems and trihalomethanes in the municipal water supplied to domestic users. Trihalomethanes are a potential carcinogen, produced by the reaction of certain disinfectants with organic materials in the water.

4.2.3 Organic Input From the Watershed

With 50% of the reservoir's contiguous watershed in forest or grassland, there is a large yearly influx of organic material into the reservoir. This contributes to the nutrient loading of the reservoir, supporting algal growth. Decomposition of the organic material depletes oxygen values in the water.

4.2.4 Low Dissolved Oxygen Levels

Both items above, algal blooms and organic influx from the forested watershed, were said to contribute to low dissolved oxygen levels. Bacterial decomposition of algal cells or vegetative matter that settles to the bottom of the reservoir is responsible for the depletion of oxygen in the water. Low dissolved oxygen levels often result in fish kills, reducing the overall value of the reservoir for recreational purposes. Additionally, after the oxygen levels have been reduced, further decomposition will take place by anaerobic decomposition processes. This often causes release of phosphorus, tied up in bottom sediments, to the reservoir water, where it can be used as a nutrient to promote additional algae growth.

4.2.5 Reduction of Aquatic Macrophytes

Although the reservoir contained numerous aquatic macrophytes when it was first filled, a 1989 survey found that there was no submersed vegetation and emergent aquatic macrophytes were found only in shallow water along the low-gradient north shore. A reduction of water transparency, due to suspended sediment and more importantly due to thick, persistent algae blooms is responsible for the deterioration in numbers of aquatic macrophytes.

Aquatic macrophytes are important to the overall health of a reservoir ecosystem for several reasons. They provide habitat and food supplies for young fish and invertebrates. They can provide some protection for shorelines, reducing wave energy and subsequent erosion of shorelines. Additionally, macrophytes will compete with algae for common nutrients, which could potentially reduce the size and duration of algae blooms.

5. DISCUSSION

Charleston has clearly benefitted from building a reservoir that has a greater capacity and potentially a longer useful life. Water quality, at least from a sedimentological and chemical point of view has been improved. The concept of a dual-watershed has allowed the city to exercise greater control over the water that is introduced into the reservoir. However, a new set of problems has surfaced in the decade since the reservoir was built. The city has been working with personnel from Eastern Illinois University, representatives from the state and federal EPA and the Illinois State Water Survey to investigate possible solutions to the current problems. In 1989 the city was awarded a grant from the USEPA under Section 314 of the Clean Water Act. The grant provided funding for the Phase 1: Diagnostic/Feasibility Study of the side-channel reservoir (Phase 2 deals with restoration and Phase 3 is long-term monitoring); the city has applied for a Phase 2 grant.

Proposals for mitigation of sedimentation problems include land use zoning within the reservoir watershed; the ordinance is in place. Sedimentation basins are proposed for the small tributaries that directly enter the reservoir; the basins are designed so they may easily be emptied when they fill with sediment. Additionally, farmers within the reservoir watershed are adopting no-till farming practices to minimize erosion.

Shoreline erosion is a more difficult problem to solve. It has been determined that 39% of the shoreline is suffering from high erosion, 27% moderate erosion, while 34% is non-erodible. Those areas nearest to boat ramps can be protected with additional concrete aprons or rip-rap, however those areas with the highest amounts of erosion are also the most inaccessible and the hardest to treat. The city has taken the position that it will try to minimize shoreline erosion with as many non-constructive techniques as possible. Currently there are several test plots in place where dormant willow cuttings were installed in the spring of 1992. By summer of 1992 it was determined that 94% of the cuttings survived and were growing. Other species will be introduced in spring of 1993, they include: Baldcypress, Cottonwood, Sycamore, Willow Oak and Water Oak. If the bio-remediation techniques prove successful the shoreline will be more aesthetically pleasing and more sensitive to the nurturing of wildlife populations.

Algal blooms have been treated with conventional means, by using copper sulfate when needed. It is doubtful whether the problem can be eliminated. However, it can be minimized by decreasing the amount of sediment introduced into the reservoir from tributaries and shoreline erosion. This will decrease nutrient loading which promotes algae growth. Once shorelines are stabilized, hopefully with vegetation, emergent and later submersed aquatic macrophytes can be introduced that will filter sediment and compete with algae for the available nutrients. Pre-treatment of water pumped from the river to the reservoir would decrease the total amount of sediment, pesticides, herbicides, nutrients, and algae introduced to the reservoir from the Embarras River watershed.

At this point in time little can be done about low dissolved oxygen levels in the reservoir. A reduction of algae and organics from the watershed should help. The city has installed aerators in a five-acre area surrounding the intakes where water is pumped from the reservoir to the water treatment plant. Although this does little for the dissolved oxygen levels in the reservoir as a whole, it does provide oxygenated raw water to the water plant, allowing preferred methods of treatment for potable water for domestic consumption.

Thus, although the city has benefitted from an improvement in water quantity and quality, new problems have surfaced which may cause future degradation of the municipal raw water supply. A basic problem is that the side-channel reservoir, as designed, has no through-flow capability, acting as a sink for nutrient-rich organic material and sediment. Fortunately problems have been identified and research has begun on several remediation methods.

6. REFERENCES

1. Yang, C. T., 1974. "Sedimentation of Lake Charleston, Charleston Illinois." Illinois State Water Survey, Open File Report, Urbana: 14 p.

2. Singh, K. P., Durgunoglu, A., and Broeren, S., 1992. "Desirable, Conservation-Oriented Reservoir Design and Operation." Transactions of the Illinois State Academy of Science 85: 211-219.

3. City of Charleston, 1992, "Clean Lakes Program, Phase 1 Diagnostic/Feasibility study of the Charleston Side-Channel Reservoir." City of Charleston, Open File Report, Charleston: 203 p.

Location Decisions for Franchisees and Franchisors

B. T. Cullen and R. L. Curtiss
Department of Geography
University of New Mexico

1. INTRODUCTION

John Naisbitt, author of Megatrends, says: "Franchising is the most successful marketing concept ever created." (8) The statistics bear him out. Between 1980 and 1990, the number of franchised establishments in the U.S. increased from 442,400 to 533,000, and sales increased from $336,200 to $716,400 billion. (5) Over one third of all U.S. retail sales now take place in franchised establishments. (10) They have become ubiquitous features on the retail landscape. Familiar names such as McDonald's, 7-Eleven, Super 8 Motels, and Midas Muffler adorn almost every shopping mall, retail strip, and highway intersection in the U.S. and Canada. In a one mile stretch of Central Avenue in Albuquerque, across the street from the University of New Mexico, there are 13 different franchises. In the city of Albuquerque, over 1,000 different franchise names dot the landscape. Franchising is flourishing.

The discontinue rate for franchised businesses has been less than 5 percent every year since 1974, while small businesses in general have had a 30 to 65 percent failure rate. (8) However, simply purchasing a franchise is not a guarantee of success. Many comedy club franchises, for example, have not succeeded because their markets are saturated, talented comics are in short supply, and humor tastes vary regionally. (11) Other franchises suffer from poor site selection. The profitability of most franchises is influenced by location. Although some factors, such as site and situation, apply generally to business outlets, others, such as territorial considerations and agreements between franchisees and franchisors, are fairly unique to franchising. This article identifies and explains the types of location factors that should be considered before purchasing a franchise.

2. BACKGROUND

Franchising is a broad business format which encompasses a wide variety of business arrangements. In legal terms, it is a contractual method for marketing and distributing the goods and services of a company (the franchisor) through a dedicated or restricted network of distributors (franchisees). (3) Although a variety of contractual arrangements exist, they generally can be grouped into the following two categories: product/trade name franchises and business format franchises. (3) Product/trade name franchises essentially arrange to distribute the dealer's product, but they are not totally identified with the manufacturer and/or supplier even though they may use the dealer's trade name and logo. (9) Examples of these types of franchises include car dealerships, soft drink bottlers, and gasoline stations. Until the Arab oil embargo and subsequent restructuring of the oil industry, 48.6 percent of all franchises and 80 percent of product/trade name franchises were gasoline stations. Their share has since declined to 22 percent of the franchised businesses, but this category of franchises still accounted for two-thirds of all franchise sales in 1988. (3)

Business format franchises are now the fastest growing category and account for two-thirds of all franchised businesses. (3) They differ from product/trade name franchises, because they are completely identifiable with the dealer. The franchisee purchases not only the product, trademark, and logo, but an entire system of doing business. (9) He/she is normally obliged to follow the purchased format and maintain company standards. Examples of business format franchises include fast-food restaurants, convenience stores, hotels, and real estate agencies. With this type of arrangement, the franchisee pays the franchisor a periodic royalty in addition to an initial fee. The royalty and fee usually entitle the franchisee to benefits like business set-up assistance, advertising, and site selection. (9)

3. SELECTING A LOCATION

The first thing a prospective franchisee of a location sensitive business needs to know is who selects the site. Mark Lewis, in his book How To Franchise Your Business, recommends that a company just going into franchising should: "Let the franchisee find the location." (14) Many franchisors, however, prefer to either select the sites themselves or utilize an outside consulting firm or established franchisee for site selection. Even if the franchisor or an agent for the franchisor selects the site, the franchisee should carefully evaluate its potential. What follows are some of the location factors that should be evaluated by the franchisee whether or not site selection is his/her responsibility or the franchisor's.

3.1 AREA PORTRAIT

There are thousands of franchises, so it can be difficult to choose one that will prosper. (4) Choosing the right franchise is extremely important. Not only must one's prior experience be considered, but the location and nature of all other businesses in the area must be scrutinized. Information on potential franchises can be obtained from a directory like The 1993 Franchise Annual. (4) This directory lists available franchises and provides basic information on price, operating format, basic good or service, and number in operation.

Once one or more possible franchises are chosen, the potential franchisee should analyze market saturation. A preliminary check can be made by looking in the telephone book and identifying the locations of similar existing franchises and independent businesses. These locations can be plotted on a map for future reference.

An on-site survey is necessary after the franchisee has settled on several possible locations using the map and directories. A general feel for the area can be obtained and vacant real estate or lots can be plotted on the map. Existing businesspeople can supply information on leasing arrangements, zoning, and whether or not they are happy with their locations. According to Finn, being happy with the location is a prerequisite. (6) If the franchisee feels uncomfortable in the area, it would be foolish to purchase a franchise there.

These steps can help towards deciding on the type of franchise and where it should be located. The information collected, coupled with information amassed before the final decision, should ensure that an obvious locational mistake is not made.

3.2 TERRITORY

With the exception of the franchisor's name, probably the most valuable item purchased from the franchisor is the territory. (23) The territory is the area or region in which the franchisee may or must operate his/her facility. Territorial considerations have been important from the very beginning of franchising. In 1851, Issac Singer, perhaps the first franchisor, accepted fees from independent salesmen for territorial rights to market his sewing machine. (3)

Not all franchisors sell territorial rights, and those that do define the rights in different ways. Table 1 summarizes the findings of a study conducted by The Conference Board, Inc. (22) The results are based on information obtained about 189 franchisors. Even though the data are dated, they provide useful information on the methods franchisors use to define territorial rights.

Table 1
BASIS FOR ESTABLISHING FRANCHISE TERRITORIES

Basis For Establishing Territories	Types Of Franchises				
	Total	Fast-Food Beverage	Nonfood	Personal Services	Business Products Services
Geography only	18.5%	26.6%	12.0%	9.6%	26.4%
Population only	15.2	11.6	22.0	12.9	14.7
Combination of the two categories	55.7	61.6	50.0	66.1	41.1
Other	10.4	0.0	16.0	11.2	17.6

Source: (22)

Delimiting the physical boundaries of a region (referred to as geography in Table 1) is one method used by franchisors, especially those marketing fast-food and beverages, to define territory. Each franchisee is given the right to service an area covering a specified number of blocks, counties, square miles, etc. Problems can arise when there is not sufficient buying power in the territory to support the facility, when there is a shift in the population distribution, when the composition of the population is not compatible with the facility, or when there are too many established competitors. This is what Webster calls the "Population Pitfall". (23) In other words, the franchisee should conduct a trade area analysis before signing any contractual agreement.

Other franchisors, mainly nonfood retailers, base territory strictly on population (see Table 1). Many of the same pitfalls mentioned above can affect franchisees who fail to evaluate their potential market. An additional problem exists in that several franchisees of the same company might be directly competing with each other, because they are located too close together. This results because the franchisor has determined how many facilities a local market can support and sold that number of franchises without specifying where within the market they would locate.

Most companies, however, use elements of both population and spatial extent when defining their territories. The exact procedure is often detailed in the franchise agreement. The franchisee needs to research what the phrase "exclusive territory" in the contractual agreement with the franchisor implies. Some of the questions that the franchisee needs to have answered are: (10,23)

-Does the exclusive right apply for only a specified period of time?

-Does the franchisor have the right to "buy-back" the franchise and territory when the existing contract expires?

-Can future franchisees locate right on the existing border, pirating away part of the trade territory?

-Do existing franchisees have the right of first refusal

on new franchises in bordering territories?

- Does the franchisor have the right to open a "model" or "demonstration facility" in the exclusive territory?

- Is the territory contingent upon the royalties qenerated?

- Can the franchise be relocated if the lease expires, the trade area deteriorates, or the property is condemned?

If the franchisor does not sell exclusive territories, the franchisee should determine if there are limitations on what other franchisees or the franchisor can do in the territory. The last thing a franchisee wants is competition from facilities of the same company and especially the franchisor.

3.3 TRADE AREA ANALYSIS

Trade area analyses are often easier for franchisees than independent businesspeople. Large franchisors have often done intensive research on the types of individuals who frequent their facilities or utilize their service. (23,21) Good franchisors have often already conducted trade area analyses, selecting those territories where they want to locate franchises. In some cases, sites are stockpiled for future expansion. (23) If customer profiles are available from the franchisor, the franchisee should be able to verify, using information from the Bureau of the Census, local banks, and the Chamber of Commerce, whether or not the demographic and socioeconomic characteristics of the proposed territory are compatible with the franchise. If customer profiles are not available from the franchisor, they can often be developed by talking with successful franchisees and analyzing the demographic and socioeconomic characteristics of their territories.

If the above mentioned customer profiles are not available or easily obtainable, most authors agree that the franchisee and/or franchisor should, at a minimum, evaluate the following characteristics of the territory: (6,7,14,16,21,23)

- Population: analyze density, growth rates, and growth potential

- Demographics: determine age of population by sex, household size, and household relationship of all members of the household

- Socioeconomic: evaluate the ethnicity, occupation of household head, education levels, income, and automobile and home ownership

- Character: access quality of commercial and residential activity and future development plans

3.4 SITE SELECTION

Site selection will be strongly influenced by the nature of the franchise. Self-generating franchises, such as discount houses, automotive stores, and drive-in-theaters, attract people to their location and are, therefore, less concerned with finding the optimum location. (16) In contrast, businesses dependent on transient trade must find a site with maximum accessibility. Agglomerating with related businesses is critical for franchises that sell products that are easily differentiated. For example, comparative shopping usually characterizes consumer selection of a car, pair of shoes, sofa, or wedding gown. (16)

According to Webster, "helping a potential franchisee find the best possible location at the best possible price, along with building a store or fixing up a storefront, are among any franchisor's key responsibilities in the franchisor-franchisee relationship." (23)

Many large franchisors utilize data from existing locations to develop analog and/or regression models for predicting sales at potential new locations. Sometimes separate models are developed for rural areas, established strip malls, the CBD, etc. (12) In many cases, the franchisor purchases the site and builds upon it before the franchisee is selected. The franchisee then leases the premises from the franchisor. Over a third of the franchisors are heavily involved in the real estate of their franchisees. (23)

If models have not been developed, and site selection is left to the franchisee, the following factors should be evaluated before a lease is signed or a lot purchased: (6,7,21,23)

-Accessibility: corner lot, number of traffic lanes, medians or dividers, speed limit, volume of auto and pedestrian traffic, and parking spaces

-Neighborhood: types of land uses, types of dwelling units, composition of existing facilities, and visibility of store front

-Facility: purchase or lease premises, renewability of lease, specifications for the physical layout of the building and the grounds, costs of improvements, and financing

4. ADVANTAGES AND DISADVANTAGES

4.1 FRANCHISE ADVANTAGES

There are several advantages to purchasing a franchise rather than starting a new, independent business. The first advantage is the experience of the franchisor. Time and money are saved when a proven product and method of doing business is purchased, because it eliminates experimenting and fine tuning the product and/or format. Many franchisors also provide training for new franchisees, which is a second advantage. A third advantage is the economies of scale associated with multiple facility firms. Materials can be purchased in bulk and shared by the individual franchises. Advertising costs can also be shared, and, as the number of franchises increases, name recognition can translate into increased sales. Large franchises also have the capital and labor required for research and development. New products can be developed and tested before they are offered to the public. The final advantage is the assistance that can be provided by both the franchisor and other franchisees. Business synergy can be created. (10)

4.2 FRANCHISE DISADVANTAGES

But the system is not perfect. There are disadvantages with franchising and challenges for the franchisees. An important characteristic that distinguishes a franchisee from an independent businessperson is the loss of independence. Franchisees are expected to follow the directions of the franchisor and work within the system. They depend heavily on the franchisor for support. The franchisee's and franchisor's relationship is, therefore, extremely important for both parties.

The relationship becomes clouded, however, when the franchisor is in financial trouble or new owners take control. The recent financial crisis at Nutri/System, for example, threatens the franchisees' supply of an exclusive line of die food, which is a big source of the franchisees' income. (19) The situation creates uncertainty, because the franchisees do not know how the crisis will be resolved. The franchisees are optimistic, however, that new owners will correct the situation. But new ownership is not always a blessing for the franchisees. When Service Master's TruGreen subsidiary boughtout ChemLawn Services Corp., some franchisees found themselves competing against their new franchisor. (20)

Relationships are also strained when franchisees are in direct competition with each other. This can happen when a franchisor over-saturates a market area. For example, when McDonald's Corp. opens a new restaurant, business is sometimes diverted from existing McDonald's outlets. Existing contracts with McDonald's Corp. do not provide the franchisees with protected territories, any right of first refusal, or compensation for business allegedly diverted. (18)

Growth is the principal financial goal of some franchisors. The franchisor's royalties and fees increase as the number of outlets increase whether or not the individual outlets are profitable. Some Subway Sandwich Shop's franchisees, for example, feel that the chain is expanding so rapidly that sites are sometimes poorly selected and too close together, making some of the outlets unprofitable. The number of Subway franchises increased from 2,300 in 1987 to nearly 7,000 in 1992. (15,1) According to an article in The Wall Street Journal, little time is devoted to site selection. (15)

Profit is not ensured just because the franchisor evaluated the site and situation. An AlphaGraphics franchisee, for example, did not question the franchisor when he was told that there were only 11 competitors within four blocks of his proposed site. But the franchisor's research only included print shops visible from the street. Actually, there were 22 existing print shops in his service area, when print shops in the upper floors of office buildings were included. (17)

Even though franchising lowers the risk of business failure, it does not preclude it. Abuses by some franchisors are well documented and increasing resulting in legislation which more clearly specifies the rights of the franchisees. (2,18)

5. CONCLUSIONS

By the turn of the century, franchising will be the "primary method of doing business" in the U.S. (3) The advantages of franchising include the experience and training provided by the franchisor, economies of scale, and business synergy. Even though the advantages outweigh the loss of independence that results when an individual purchases a franchise, the franchisee still needs to carefully research the company, product line, contractual arrangement, and potential locations before making a commitment. A poor location decision, for example, will reduce the profitability of the business. As has been stressed in this paper, the franchisee should carefully analyze how franchise territories are established, whether the trade area is compatible with the franchise, and the relative location, accessibility, and financing of the specific site. If a prize location is selected, the franchisee must be wary of "the buy-back" by the franchisor. Franchising is the wave of the future, but not risk free.

6. REFERENCES

1. Anton, M. and E. Wallace, 1993. "Understanding The Franchise 500." Entrepreneur. January: 130-227.

2. Bahls, J.E., 1993. "Franchisee Vs. Franchisor." Entrepreneur. January: 83-85.

3. Committee On Small Business, House Of Representatives, 1990. Franchising In The U.S. Economy: Prospects and Problems. Washington, D.C.: U.S. Government Printing Office, 101st Congress, Second Session.

4. Dixon, Ted (ed.), 1993. The 1993 Franchise Annual. St. Catharines, Ontario: INFO Press, Inc.

5. Economic and Statistics Administration, Bureau of the Census, 1992. Statistical Abstract of the United States. Lanham, Maryland: Bernan Press.

6. Finn, R.P., 1979. "Choosing a Good Location." Your Fortune In Franchises. Chicago: Contemporary Books, Inc., 39-45.

7. Frost and Sullivan, 1976. "Site/Location Analysis." The Franchising Market. New York: 372-374.

8. Huber, J., 1993. "Franchise Forecast." Entrepreneur. January: 72-75.

9. International Franchise Association, 1987. What you need to know when you buy a franchise. Washington, D.C.

10. International Franchise Association, 1990. Franchise Opportunities Guide. Washington, D.C.

11. Jefferson, D.J., 1992. "Comedy Clubs Aren't Big Hits As Franchises." The Wall Street Journal. January 9: B1.

12. Jones, K. and J. Simmons, 1987. Location, Location, Location. Analyzing The Retail Environment. Toronto: Methuen Publications.

13. Kinch, J.E. and J.P. Haynes, 1986. "Site selection checklist." Franchising: The Inside Story. Wilmington, Del.: Tri-Mark Publishing Co., Inc., 193-195.

14. Lewis, M.O., 1981. "Simplified Site Selection." How To Franchise Your Business. New York: Pilot Books, 10.

15. Marsh, B., 1992. "Franchise Realities: Sandwich-Shop Chain Surges, but to Run One Can Take Heroic Effort." The Wall Stree Journal. September 16: A1, A10.

16. Seltz, D.D., 1980. "How To Select A Proper Location For Your Franchised Business." How To Get Started In Your Own Franchised Business. New York: Farnsworth Publishing Company, Inc., 111-117.

17. Selz, M., 1992. "Many Ex-Executives Turn to Franchising, Often Find Frustration." The Wall Street Journal. October 14: A1, A5.

18. Tannenbaum, J.A., 1992. "Curbing Franchising Abuses; Guide Comes Under Fire." The Wall Street Journal. May 22: B2.

19. Tannenbaum, J.A. and L.L. Valeriano, 1993. "Nutri/System Franchisees Live Franchiser's Nightmare." The Wall Street Journal. May 3: B2.

20. Tannenbaum, J.A., 1992. "Regulators Press for Wider Disclosure by Franchisers." The Wall Street Journal. August 13: B2.

21. Tarbutton, L.T., 1986. "The Site Approval Committee." Franchising: The How-To Book. Englewood Cliffs, N.J.: Prentice Hall, 128-136.

22. The Conference Board, Inc., 1971. "Site Selection Practices." Franchised Distribution. New York: 71-74.

23. Webster, B., 1986. "Location, Location, Location." The Insider's Guide To Franchising. New York: American Management Association, 70-94.

RESOLUTION-DEPENDENT EFFECTS ON CLASSIFICATION ACCURACY IN REMOTE SENSING

Howard Veregin and Ferdouse Sultana
Department of Geography
Kent State University
Kent, Ohio, 44242-0001

1. ABSTRACT

This study addresses the issue of land cover classification accuracy for remotely sensed imagery. Most classification algorithms assume that a pixel contains only one land cover class. This can lead to inaccuracies when there are mixed pixels containing more than one class. The fuzzy c-means classifier has been shown to be useful for estimating the proportions of different classes within mixed pixels. However, results appear to be dependent on the degree of spatial variability in spectral response associated with the class under consideration. Spatial variability interacts with sensor spatial resolution, such that an increase in resolution may have either a positive or a negative effect on classification accuracy. This study examines the effects of variations in spatial resolution and spatial variability on the results of the fuzzy c-means classifier. Landsat Thematic Mapper imagery of Kent, Ohio, is used in the study.

2. CLASSIFICATION ACCURACY

Users of remotely sensed imagery are often faced with the problem of assessing land cover classification accuracy. Image classification is achieved through the assignment of pixels into land cover classes based on pixel spectral responses in one or more bands of the electro-magnetic spectrum. Classification accuracy is affected by the level of spatial resolution achievable by the sensing system. For the multi-spectral scanning systems used on environmental monitoring satellites, resolution is defined in terms of sensor instantaneous field of view (IFOV), an angular measurement defining the ground area represented by each pixel in the digital database. In general it can be expected that as spatial resolution is enhanced (i.e., as pixel size declines), classification accuracy will improve due to the ability of the sensing system to identify more detail on the ground.

This relationship is complicated by spatial variability in spectral response associated with a given land cover class. Spatial variability refers to the degree to which ground features with different spectral responses are dispersed over space. Certain land cover classes contain a heterogeneous mixture of constituent classes exhibiting significant variations in spectral response. For example, the urban class exhibits high-frequency spatial variation in spectral response as it contains numerous constituent classes (e.g., buildings, vegetation, transportation features) for which the spectral response varies significantly over even relatively short distances. Other classes (e.g., agricultural) exhibit low frequency spatial variation in spectral response, since spectral response is relatively constant even over great distances.

Classification accuracy is affected by the interaction between spatial variability and sensor spatial resolution. This is illustrated in Figure 1. The figure shows an area of urban land cover at two different spatial resolution levels. At the 60-meter resolution level, each pixel has an approximately equal mixture of constituent classes, while at the 30-meter resolution level, each pixel contains a different mixture of constituent classes. The 60-meter pixels have a spectral response typical of the integrated response of all constituent classes. In contrast, the spectral response associated with each of the 30-meter pixels depends on the class that dominates within the pixel. The result, as shown in the spectral space diagrams in Figure 1, is a higher degree of spectral heterogeneity for the 30-meter resolution pixels. In effect, the size of the spectral space for the urban class expands as spatial resolution is enhanced. This results in a decrease in the ability to identify clusters of pixels representing meaningful classes, and an associated decrease in classification accuracy.

In areas where spatial variability is low (e.g., agricultural areas), changes in spatial resolution have the opposite effect on classification accuracy. If constituent classes are few in number or do not exhibit significant differences in spectral response, then there is no expansion in the size of the spectral space as resolution is enhanced. Instead, classification accuracy improves due to a decline in the number of mixed pixels. Mixed pixels are pixels containing more than one land cover class. Mixed pixels are common along the boundary between two classes, and also tend to occur for classes that contain a mixture of constituent classes (e.g., urban). Classification of mixed pixels into a single class inevitably leads to information loss and inaccuracy in land cover maps.

3. SPECTRAL HETEROGENEITY AND MIXED PIXELS

The degree of spectral heterogeneity and the frequency of mixed pixels tend to have counteracting effects on classification accuracy. (6) In particular, an enhancement in spatial resolution results in

(1) an increase in spectral heterogeneity, which is associated with a decrease in classification accuracy, and

(2) a decrease in the number of mixed pixels, which is associated with an increase in classification accuracy.

The relative importance of these two factors for a given land cover class depends on the level of spatial variability in spectral response for that class. For classes with high spatial variability, the effects of spectral heterogeneity dominate, and accuracy tends to decrease with an enhancement in spatial resolution. Spectral heterogeneity is averaged out at coarser resolution levels, because all pixels of a particular class have approximately the same mixture of constituent classes. This causes a reduction in the size of the spectral space for the class and less overlap with other classes. Classification accuracy is therefore often higher at coarser spatial resolution levels for areas of high spatial variability. (8)

Previous research shows that different land cover classes are associated with different levels of spectral heterogeneity, that different spectral bands can produce different levels of spectral heterogeneity for the same classes, and that the change in spectral heterogeneity due to a coarsening of spatial resolution depends on the land cover class and spectral band. (6) Other research indicates that, the greater the degree of spectral heterogeneity, the greater the increase in accuracy resulting from a coarsening of spatial resolution. (2)

For areas of low spatial variability, the mixed pixel effect tends to dominate. Classification accuracy therefore tends to increase with an enhancement in spatial resolution, since the number of mixed pixels decreases as resolution is enhanced. Thus in areas of low spatial variability, an enhancement in the level of resolution will result in improved classification accuracy.

4. FUZZY CLASSIFICATION

Traditional classification algorithms assume that the spectral response of a given pixel represents only one land cover class. The presence of mixed pixels contradicts this assumption. The existence of mixed pixels is explicitly accounted for in classification algorithms based on fuzzy set theory. Fuzzy set theory recognizes the inherent spatial variability in geographical phenomena in the form of classes that do not possess precise boundaries.

The fuzzy c-means classifier allows such classes to be derived from remotely sensed data. (1) The classifier allows for estimation of the probabilities of membership of multiple classes within pixels. These probabilities are referred to as membership function values. Empirical evidence shows that membership function values are often highly correlated with the relative area of each class within pixels. (3,4,5) However, the strength of the correlation varies among classes, with lower correlations for classes exhibiting higher spatial variability in spectral response. Correlation coefficients have been observed to be on the order of 0.9 for homogeneous classes (e.g., water and wetlands) and on the order of 0.6 for more heterogeneous classes (e.g., buildings and open land).

The fuzzy c-means classifier can be used to estimate the proportions of component classes in a pixel. Results suggest that this ability is affected by spatial variability. (3,4,5) However, this issue has not been addressed explicitly in previous research. Since spatial resolution affects the nature of the spectral response recorded as a function of spatial variability, it is likely that the relationship between membership function values and cover class proportions will vary as a function of the interaction between spatial resolution and spatial variability.

5. OBJECTIVES AND METHODOLOGY

The objective of this study is to examine the effects of changes in spatial resolution on the accuracy of the fuzzy c-means classifier. The specific hypothesis to be tested is that the correlation between the membership function values derived from the classifier and the actual proportion of pixel area covered by each land cover class varies as a function of sensor spatial resolution and the level of spatial variability. In particular, following degradation in the level of spatial resolution, the correlation between membership function values and actual land cover class proportions will increase in areas of high spatial variability and decrease in areas of low spatial variability.

The study area is centered on the city of Kent, Ohio. The area includes land cover classes with different levels of spatial variability. A 120-row by 160-column subset of a Landsat TM (Thematic Mapper) image for April 10, 1987 was extracted for the study area. All seven bands of the TM image were used. The image has a spatial resolution of 30 meters.

Data processing and analysis was carried out in four stages.

First, the original 30-meter resolution was degraded to produce images at two coarser resolutions of 60 meters and 120 meters. These images are 60 rows by 80 columns, and 30 rows by 40 columns, respectively. Resolution degradation was achieved by aggregating blocks of four adjacent pixels from the next-highest resolution level.

The fuzzy c-means classifier was then applied to obtain fuzzy membership values for the land cover classes. The classifier was translated from FORTRAN into C and implemented on a UNIX workstation. The original FORTRAN version can be found in Bezdek et al. (1) The classifier is based on a non-hierarchical clustering algorithm that measures pixel composition by first randomly assigning pixels to a land cover class. The classifier uses the least squares method to minimize the difference between within-class variation, which determines an optimal clustering of land cover classes. (1,3,4,5,7)

The fuzzy c-means classifier was applied independently to each resolution level of TM imagery. Six information classes were originally extracted. Several of these classes were then collapsed to produce four information classes labeled "urban", "vegetation", "water", and "other". Figure 2 shows the classification results at each resolution level. This figure was produced by assigning to each pixel the class value with the highest membership function value.

Estimates of pixel area covered by each land cover class were derived for a random sample of pixels. Twenty-five pixels were selected independently for each resolution level. Class areas were estimated using dot planimetry and converted to proportions of total pixel area. A series of 1:1200 engineering maps prepared in 1982 for the City of Kent Department of Engineering were used for this purpose.

The relationships between fuzzy membership values and area proportions for each class were assessed using correlation and regression analysis. Correlation analysis was used to assess the strength of the relationships, while regression analysis was used to assess the degree to which fuzzy membership values and area proportions exhibit a one-to-one correspondence.

6. RESULTS

Table 1 shows the correlations between fuzzy membership values and area proportions for each class and spatial resolution level. The correlation coefficient for the "urban" class increases from 0.2320 to 0.3208 as resolution is degraded from 30 to 120 meters. The opposite trend is evident for the "water" class. In this case, the correlation coefficient decreases from 0.7913 to 0.3278 as resolution is degraded. These results support the hypothesis that, following degradation in the level of spatial resolution, the correlation will increase in areas of high spatial variability (i.e., urban), and decrease in areas of low spatial variability (i.e., water).

An increase in the correlation coefficient (from -0.1089 to 0.3984) as resolution is degraded is evident for the "other" class. This would suggest that this class contains a mixture of constituent classes (e.g., agricultural land with vegetation at various stages of development) that exhibit significant differences in spectral response.

No clear trend is evident for the "vegetation" class. The correlation varies from a value of -0.0806 at 30-meter resolution to -0.2521 at 120-meter resolution. The negative signs for these coefficients suggest that there is disagreement between the membership function values and area proportions for this class. Indeed the membership values for vegetation seem to correlate with the proportions for the "urban" and "other" classes, while the proportions for the "vegetation" class correlate best with the membership values for the "other" class. The cause of these discrepancies is unclear, but may be due to the five-year difference between the dates of the image and engineering maps.

Table 2 shows the results of the regression analysis between the fuzzy membership values and area proportions. In order for the membership values to be used directly to compute the proportions, the intercept and slope coefficients should not be statistically significantly different from 0 and 1, respectively. This would indicate a one-to-one correspondence between the two sets of values. As Table 2 shows, however, most intercept coefficients are significantly different from 0 and most slope coefficients are significantly different from 1. There does not appear to be any systematic change in the level of significance as spatial resolution levels change.

7. CONCLUSIONS

From the results of this study the following conclusions can be reached.

(1) There are significant correlations between fuzzy membership values and area proportions for most land cover classes.

(2) The strength of the correlation varies among classes depending on spatial variability, as can be seen in the contrasting values for the "urban" and "water" classes.

(3) As spatial resolution is degraded, the strength of the correlation increases in areas of high spatial variability (e.g., urban) and decreases in areas of low spatial variability (e.g., water).

(4) There is not a one-to-one correspondence between fuzzy membership values and area proportions, which indicates that membership values cannot be used directly as surrogate estimates of these proportions.

8. REFERENCES

1. Bezdek, J. C., Ehrlich, R., & Full, W. 1984. FCM: The fuzzy c-means clustering algorithm, Computers and Geosciences 10: 191-203.

2. Cushnie, J. L. 1987. The Interactive Effect of Spatial Resolution and Degree of Internal Variability Within Land Cover Types on Classification Accuracies. International Journal of Remote Sensing 8: 15-29.

3. Fisher, P. F., & Pathirana, S. 1989. Urban Boundary Detection From Landsat Imagery: A GIS and Knowledge-Based Approach to MSS Data. Proceedings of the ASPRS/ACSM Annual Meeting 4: 93-101.

4. Fisher, P. F., & Pathirana, S. 1989. Evaluation of Fuzzy Membership of Land Cover Classes in Suburban Areas of North-east Ohio. Proceedings of the ASPRS/ACSM Fall Convention: 123-132.

5. Fisher, P. F., & Pathirana, S. 1990. The Evaluation of Fuzzy Membership of Land Cover Classes in the Suburban Zone. Remote Sensing of Environment 32.

6. Markham, B. L., & Townshend, J. R. G. 1981. Land Cover Classification Accuracy As A Function of Sensor Spatial Resolution. Fifteenth International Symposium on Remote Sensing of Environment: 1075-1090.

7. Robinson, V. B., & Thongs, D. 1985. Fuzzy Set Theory Applied to the Mixed Pixel Problem of Multispectral Land Cover Databases, in B. Opitz (ed.), Geographic Information Systems in Government, pp. 871-886.

8. Toll, D. L. 1985. Effect of Landsat Thematic Mapper Sensor Parameters on Land Cover Classification. Remote Sensing of Environment 17: 129-140.

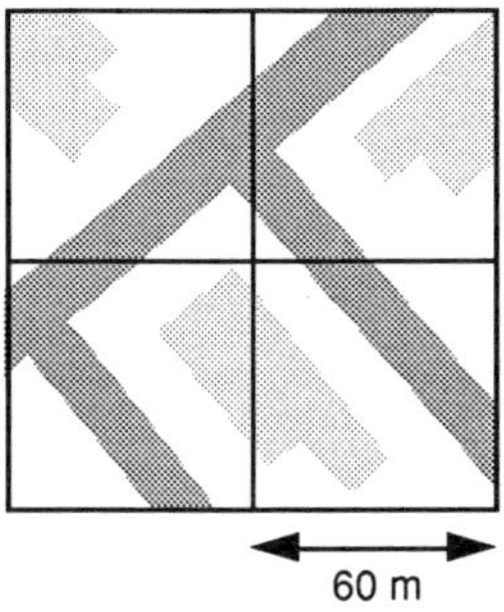

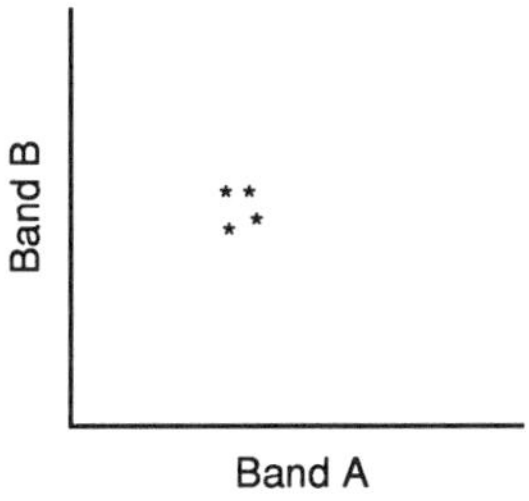

At 60-meter resolution, each pixel has an equal mixture of constituent classes (buildings, transportation and vegetation).

In spectral space, the 60-meter pixels have an integrated response from the three constituent classes.

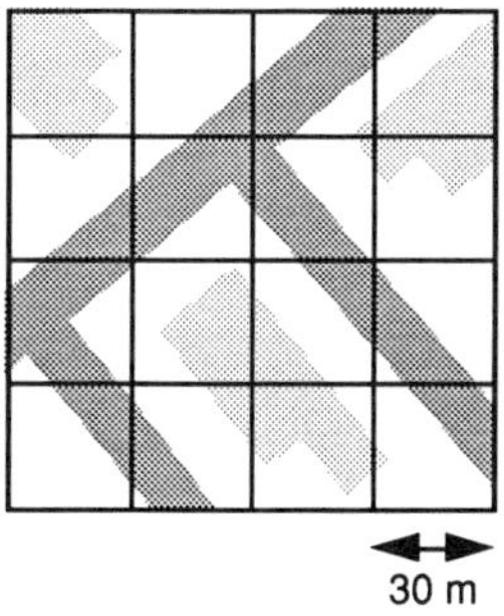

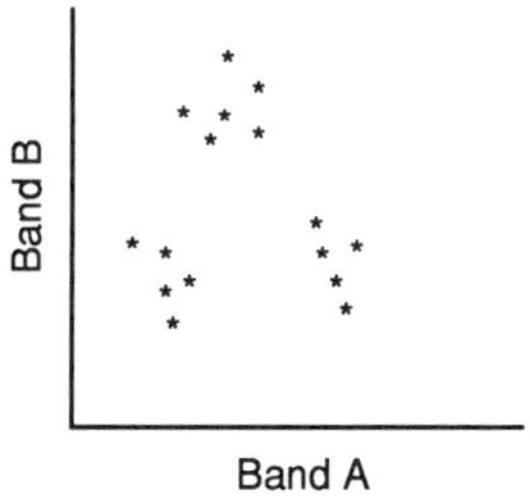

At 30-meter resolution, each pixel has a different mixture of constituent classes.

In spectral space, the 30-meter pixels exhibit a high degree of spectral heterogeneity.

Figure 1. Relationship between Spatial Resolution and Spectral Heterogeneity

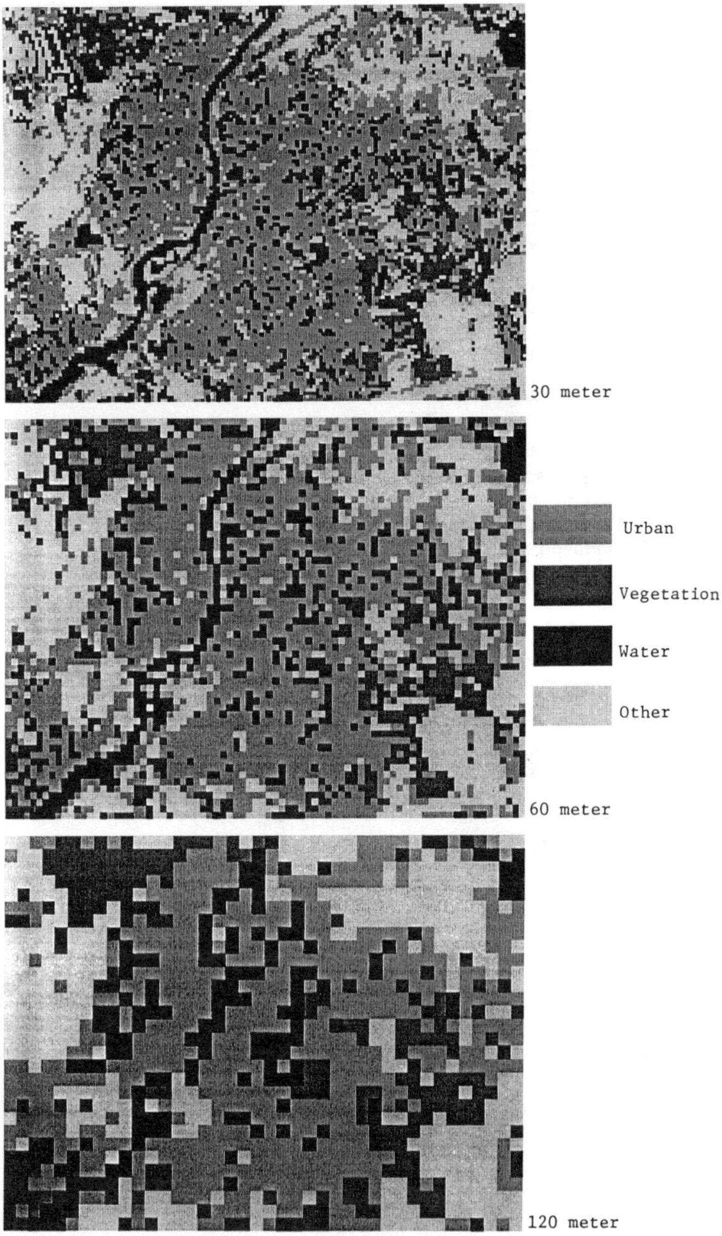

Figure 2. Classification Results Showing Class with Maximum Membership Function Value

Class Proportions		Fuzzy Membership Values: Urban	Vegetation	Water	Other
Urban	(30 m)	**0.2320**	0.2531	0.1348	-0.1488
	(60 m)	**0.2581**	0.0918	0.1363	-0.4785***
	(120 m)	**0.3208***	0.0682	-0.0229	-0.4474***
Vegetation	(30 m)	-0.2335	**-0.0806**	-0.1000	0.3548**
	(60 m)	-0.0901	**-0.2092**	-0.1926	0.4117***
	(120 m)	-0.1700	**-0.2521**	0.2741*	0.2111
Water	(30 m)	-0.1721	-0.1430	**0.7913*****	-0.1709
	(60 m)	-0.1336	-0.1069	**0.3242***	-0.1189
	(120 m)	-0.1336	-0.0684	**0.3278***	-0.0249
Other	(30 m)	-0.0964	0.3951**	-0.0984	**-0.1089**
	(60 m)	-0.2133	0.0836	-0.1189	**0.3135***
	(120 m)	-0.2380	0.1209	-0.2510	**0.3984*****

* $p \leq 0.25$
** $p \leq 0.10$
*** $p \leq 0.05$

Table 1. Correlation Results

	Class	Intercept	Slope
Urban	(30 m)	0.250	0.435*
	(60 m)	0.370***	0.366***
	(120 m)	0.265**	0.403***
Vegetation	(30 m)	0.246***	-0.177***
	(60 m)	0.275***	-0.379***
	(120 m)	0.276***	-0.317***
Water	(30 m)	-0.006***	0.099***
	(60 m)	0.004***	0.124***
	(120 m)	0.001	0.119***
Other	(30 m)	0.366***	-0.175***
	(60 m)	0.118*	0.395***
	(120 m)	0.188***	0.534**

* $p \leq 0.25$
** $p \leq 0.10$
*** $p \leq 0.05$

Table 2. Regression Results

DELINEATION OF THE PROPOSED OHIO & ERIE CANAL NATIONAL HERITAGE CORRIDOR USING A GEOGRAPHIC INFORMATION SYSTEM

Laura L. Clark
Department of Geography
Kent State University
Kent, OH 44242

1. INTRODUCTION

The corridor containing the historic Ohio & Erie Canal is currently being studied to determine the suitability and feasibility of adding it to the National Park System through designation as a National Heritage Corridor. A heritage corridor is a "linear landscape of significant and diverse character that becomes the focus of a public and private partnership to preserve, enhance, and interpret the cultural, natural, and recreational resources within to create a sense of place, promote economic growth, and enhance the image and quality of life in the region".(5) In brief, a corridor can be defined as an area that exhibits unifying features and relationships.(2)

While the establishment of a national heritage corridor is not as dramatic as that of a national park, wilderness area, or nature preserve it nevertheless has effects on many interests. The problem lies in how to best serve all interests and representations associated with the proposed corridor. Land ownership of actual canal parcels includes local, state, and federal governments, industry and individuals. Many diverse special interest groups have formed to support the designation of the corridor and the protection, preservation, development, and enhancement of the cultural, natural, industrial and recreational resources of the corridor. This study seeks to recommend potential boundaries which are most appropriate in benefiting industry, individuals, and government associated with the corridor and which do not adversely affect the natural and cultural resources which are to be preserved and enhanced.

The application of the boundary selection lends itself well to the collection, storage and analysis capabilities of a GIS. Once a database is established the study could be an important aid in the ongoing planning process for this corridor as well as aiding in the planning of future corridors or greenways. Utilizing a GIS would result in boundary recommendations that strive for positive relationships between the corridor and the governments, communities, and special interest groups associated with it.

To this end the study will be concerned with the unique attributes of the Ohio & Erie Canal and the corridor it defines. Thus background on the canal's history and present state will be discussed. The three existing national heritage corridors will be examined with respect to their boundaries and the criteria used in their selection. The specific criteria which are important in the boundary selection for the Ohio & Erie Canal corridor will be examined as well. The methods used in the study will be explained and the products which are expected to result will be discussed. It may be noted that this is a masters thesis work in progress and as such is subject to change.

2. BACKGROUND

2.1 HISTORY

October 15, 1992, marked the 160th anniversary of the completion of the Ohio & Erie Canal which completed the connection of Cleveland and Lake Erie to Portsmouth, Ohio and the Ohio River. The construction of the canal was the "single most important historical development in the Cuyahoga Valley".(6) Before 1825, when construction of the canal began, most of Ohio was sparsely populated, with an economy based on barter and subsistence agriculture due to the extreme hardships of travel, inclement weather, lack of quick access to markets and near absence of money.(6) The canal brought cash, commerce and development to this landscape. Completed in 1832, the canal changed Cleveland from a frontier to a bustling port city and had far reaching effects on all of northeastern Ohio.

2.2 IMPORTANCE

But its importance extends beyond its history. The Ohio & Erie Canal corridor presents an open-space connection linking parks, nature reserves, cultural features, and historic sites with each other and with populated areas. In his book, "Greenways for America", Little discusses the relatively recent movement to establish what are now referred to as "greenways".(4) Within the last generation or two we have become aware of the importance of these scenic and/or cultural resources and endeavored to create such linkages. Linkage is the core concept of greenways, which include heritage corridors, natural corridors such as river valleys and ridgelines, and cultural corridors such as canals, scenic roads and railroad right-of-ways.

2.3 PRESENT BOUNDARIES

A boundary has been set for the northern 7 miles of the proposed corridor which stretches from Lake Erie and Cleveland to the northern boundary of the Cuyahoga Valley National Recreation Area (CVNRA). The boundary in this portion encompasses the land within the Cuyahoga River Valley as well as some Cleveland neighborhoods above the rim of the Valley and areas that function as part of the Corridor.(2) The east and west boundaries of the CVNRA will define the corridor boundaries for that portion of the corridor. What remains is approximately 35 miles of canal corridor stretching from the southern boundary of the CVNRA, through Akron, to the village of Zoar, the southern terminus of the corridor.

This study will concentrate on a section of this latter portion of the canal corridor. A representative section will be chosen and boundary selection methods developed to serve as a model for remaining sections.

3. EXISTING NATIONAL HERITAGE CORRIDORS

Currently there are three National Heritage Corridors, (1) the Illinois & Michigan Canal, (2) the Blackstone River Valley and (3) the Delaware Lehigh & Navigation Canal. These were established in 1984, 1986, and 1988 respectively.

3.1 THE ILLINOIS & MICHIGAN CANAL CORRIDOR

The Illinois & Michigan Canal National Heritage Corridor is a 120 mile greenway stretching from Chicago to LaSalle/Peru. The boundaries follow major roadways and county lines. The northern boundary follows interstate highways.

3.2 THE BLACKSTONE RIVER VALLEY CORRIDOR

The Blackstone River Valley National Heritage Corridor is a 46-mile bi-state (MA and RI) corridor. The boundaries of the corridor delineate what the corridor commission defines as core cultural areas and landscapes found within the Blackstone River Valley watershed (1).

3.3 THE DELAWARE AND LEHIGH NAVIGATION CANAL CORRIDOR

The Delaware and Lehigh Navigation Canal National Heritage Corridor is 150 miles long and follows the routes of the historic Delaware Canal and the Lehigh Navigation System in eastern PA. The length of the corridor comprises four divisions, three of which are counties. The fourth is an abandoned railroad link.

4. BOUNDARY SELECTION CRITERIA

4.1 FACTORS

Factors important in the selection of the boundary include (1) cultural resources such as historic structures, historic districts and any areas which have an association with the canal, (2) natural resources such as natural areas/open space, scenic attractions, parks, wildlife, and water bodies and (3) other important factors including the viewshed from the canal, the population served, land use, land ownership and existing boundaries.

4.2 SOURCES

These factors were chosen with the help of different agencies involved in the establishment of the corridor including the CVNRA, the Ohio Office of Rivers, Trails and Conservation Assistance Program, and the Ohio & Erie Canal Corridor Coalition. Sources of data include the Ohio Department of Natural Resources, the Soil Conservation Service, the U.S. Fish and Wildlife Service, the USGS, county planning commissions, county metroparks, the Ohio Nature Conservancy, the State Historic Preservation Office, the Ohio Department of Transportation, the National Park Service, and the Office of Rivers, Trails and Conservation Assistance Programs.

5. METHODS

Because the establishment of the National Heritage Corridor is unlikely to result in any environmental costs, and because it is not to be a constructed feature with associated tangible costs, it is logical to use methods which maximize benefits rather than minimize costs.

5.1 SUITABILITY ANALYSIS

The study will yield a suitability map which shows the ability of each increment of the study area to support a given use or function, in this case the most beneficial corridor boundary. The suitability analysis approach results in a corridor boundary which maximizes benefits along the path of the boundary. This can be achieved by two different methods of suitability analysis, discussed by Hopkins.(3)

5.1.1 Linear Combination

The linear combination method to be used deals with weighted overlays. Each factor is divided into different types which are given rank values. Each rank value is then multiplied by the weight value for that factor. The factor types are mapped and overlaid upon other factor maps whereupon the weighted type values are added. This yields the suitability map, with those lengths having a higher value being the preferred boundaries. A problem with this method is that it does not deal with any dependencies that may exist between factors.

5.1.2 Rules of Combination

This method uses a set of explicit rules of combination. For example, the rating of the best factor in a given segment overrides the rating of all the other factors in that segment. In addition different sets of combinations can be rated by different rules, depending on the combination. An example of this would be that any region with a metropark within a certain distance of the canal or any tributary to the Cuyahoga River must be included within the boundaries, regardless of population served or whether or not it is within the viewshed of the canal.

In addition the rules of combination method looks at the inter-dependencies which may exist between factors. Open space will most likely be associated with parks and low population. Historic districts or areas whose development is associated with the canal will most likely be urban areas with a high population served, but lacking in wildlife, vegetation and recreation areas (as these areas are apt to be industrial). The rules of combination method has the advantage that each possible combination of factors does not have to be dealt with individually. Rules should be based in the understanding of the system which is being analyzed rather than a single set of relationships repeated for all combinations.

5.2 THE OPERATIONS RESEARCH MODEL

The second method to be used in evaluating boundaries is that of operations research, or the optimization technique. This method is based upon a model concerned with the mathematical relationships which include the objective function and constraints. The objective function is defined as the mathematical relationship representing the objective of a problem; it defines the effectiveness of the model as a function of the decision variables (or the dependent variable as a function of the independent variable). Constraints are limits or restrictions that occur in the problem. Examples of a constraint include a limit on the width of the corridor or a minimum population to be served. The goal is to find the solution to this model, the optimal location for the boundary.

5.3 PRODUCTS

The products of the study will include maps showing the results of the methods of suitability analyses and the results of the operations research model. All recommendations will be discussed. In the case that the boundary recommendations of the two methods are different, the cause for the difference will be addressed and resolved. In addition, the two methods will be compared and discussed in the final thesis.

As well as maps and the final report, a database will have been developed for the portion of the corridor under study which will prove helpful for other applications. It is hoped that this process by which an appropriate and optimal boundary was selected may serve as a model for the selection of boundaries of other corridor segments

6. REFERENCES

1. Blackstone River Valley National Heritage Corridor Commission, 1989. Cultural heritage and land management plan for the Blackstone River Valley National Heritage Corridor. Uxbridge, MA: unpublished.

2. Cuyahoga County Planning Commission, 1992. North Cuyahoga Valley Corridor Executive Summary. Cleveland, Ohio: unpublished.

3. Hopkins, Lewis D., 1977. "Methods for Generating Land Suitability Maps: A comparative Evaluation." American Institute of Planners Journal 43: 386-400.

4. Little, Charles E, 1990. Greenways for America. Baltimore, MD: The Johns Hopkins University Press.

5. Robinson, Rory, 1991. The Ohio & Erie Canal Corridor. Ohio Office of Rivers, Trails and Conservation Assistance Program, unpublished.

6. Unrau, Harlan D. and Scrattish, Nick, 1985. Historic Resource Study: Cuyahoga Valley National Recreation Area, Ohio. Denver, CO: U.S. Dept. of the Interior, National Park Service, Denver Service Center.

AN INTERACTIVE SYSTEM FOR VISUALIZATION AND EXPLORATION OF SPATIAL DATA QUALITY

Howard Veregin, Paul Krause, Robin Pandya
and Rebecca Roethlisberger
Department of Geography
Kent State University
Kent, Ohio, 44242-0001

1. INTRODUCTION

Data quality is an issue of increasing concern within the geographic information processing community. While the utility of information processing technology in support of decision-making is now widely accepted, it is increasingly being accompanied by a concern for the quality of data products, particularly when these products are used in the policy realm.(10) Due to a lack of standard methods for assessing accuracy, however, many spatial data products are still being used without a firm understanding of their reliability and limitations. There are other issues of equal importance that also remain to be solved, including the transformation of data quality characteristics as data are manipulated and transformed in the course of computer-based processing. Many cartographers, on the other hand, are more concerned with the manner in which information about data quality is communicated to the user. This communication process is essential if data quality information is to be interpreted correctly and used to guide the decision-making process. Visualization of spatial data quality forms a current research initiative of the National Center for Geographic Information and Analysis (NCGIA).(2) The issue has also been addressed by a number of researchers, although little work has been published in this area.(5,16,18)

This paper reports on the design and development of an interactive system for visualization of spatial data quality in the context of digital soil maps. The system provides tools for manipulation of spatial data displayed in different windows on a computer screen. Visual and auditory signals are used to portray variations in data quality and to permit interactive querying of the data quality component of the database. Through the use of dynamic display capabilities, the system allows for exploratory analysis of data quality. The exploratory approach permits the user to develop and test hypotheses in an informal manner. The system is designed around a flexible data quality model, which allows various dimensions of data quality to be examined.

2. DATA QUALITY FOR SOIL MAPS

A soil is an independent natural body with unique characteristics that result from interactions between climate, living matter, parent materials and relief, acting over a period of time.(17) Soil surveys -- the gathering of soil data in the field -- are compiled to produce a soil map and accompanying memoir. The soil map summarizes a large amount of information in a concise form and is convenient for use in the field. However, many users often rely on only the information found on the soil map and ignore the memoir, especially if it is lengthy or difficult to interpret.(17)

The usefulness and reliability of soil maps is dependent on the fidelity with which they depict conditions in the field. Soil maps contain inherent errors from a variety of sources, including misclassification of soil types in the field, ambiguity in the location of soil boundaries, inaccurate measurement of soil properties, interpolation error between sampled sites and cartographic inaccuracies. Map users often interpret the quality of a soil survey in terms of the reliability of the soil mapping units. The soil surveyor, in contrast, may define data quality in terms of the detail, reliability and relevance of the data that have been collected, given time and budgetary constraints. As a practical matter, a soil survey must balance quality and cost, but quality is rarely explicitly defined and is virtually never measured.(17)

Soil is classified according to soil properties, and soil map legends are based on soil classes. Mapping units are derived from soil map legends. This process necessarily translates into data that are inherently heterogeneous.(6) Compounding the problem of heterogeneity is the factor of map scale. Scale influences the choice of legend and mapping units, and affects the ability to convey complex information. The purity of a particular mapping unit is expressed as the expected percentage of random sites within the unit for which the profile class is correctly identified.(17) In the United States, mapping units must be 85 percent pure (with the exception of irrigated land), regardless of map scale. Despite this guideline, there is often little effort to verify the purity of soil mapping units.

Like most surface properties, soil varies continuously over space. Sites that are close together on the ground tend to have characteristics that are more similar than sites that are far apart. In a statistical sense sites are spatially dependent, and thus measurements made at different sites cannot be regarded as independent observations. The general method used to delineate soil boundaries involves observation of soil characteristics at specified intervals along a transect and subsequent identification of a boundary demarcating areas of different soil types. The ambiguity of soil boundaries results from the interaction between sampling frequency over space and the frequency of change in soil properties. However, soil properties do not vary uniformly over space and mapping unit boundaries are of variable width.

A variety of techniques have been used to model variations in soil boundary width. The simplest models are based on an band defining uncertainty in boundary location as a function of the soil types on either side of the boundary. Some pairs of soil types exhibit more abrupt boundaries than others, such that the width of the uncertainty band varies over the map. Other models are based on the concept of regionalized variables, fuzzy set theory, or fractal dimensionality. These models seek to model soil properties as quantities that vary continuously over space.

The usual distinction between spatial and aspatial data quality does not apply in the case of soil maps. For certain classes of geographical data, it is possible to define the quality of the spatial component (e.g., census tract boundaries) independently from the quality of the thematic component (e.g., estimates of census tract population). Soil mapping unit boundaries are dependent on the thematic component (i.e., soil properties), since these boundaries are delineated as the juncture between two different soil types. This relationship must be taken into account in the assessment and visualization of data quality for soil maps.

3. VISUALIZATION OF DATA QUALITY

In the context of this study, visualization refers to techniques that facilitate effective communication of spatial information. While visualization is usually conceived as being restricted to the visual domain, we also include other sensory domains in the definition. The discussion that follows examines both visual and auditory variables in the visualization process. The discussion of visual variables focuses on the use of color.

3.1 VISUAL VARIABLES

Traditional cartographic design is based on the application of appropriate visual variables to convey information to the map user. Bertin (4) defines seven basic visual variables -- position, size, value, texture, hue, orientation and shape. In this scheme, color is defined in terms of hue and value. Hue refers to the various colors that humans perceive (e.g., red, yellow, green). Value is the quality of lightness or darkness of a color (i.e., the amount of white or black added to a hue). Human perception of color has a third component, known as saturation. Saturation is the intensity or purity of a color. Saturation varies from zero percent (a neutral gray) to one hundred percent (maximally saturated color). Hue, value and saturation form the basis of the HSV color model often used in graphics applications.(11,12)

The predominant method of implementing color on a computer is based on the RGB (Red-Green-Blue) color model. This model employs a tristimulus representation in which a given color is described by a triple describing the relative amount of red, green and blue. These three primary colors define a Cartesian coordinate system where the subspace is a unit cube.(3) The primaries are additive in the sense that the individual contributions of each primary are added together to obtain a resulting color. Each color point within the bounds of the cube is represented as a triple where the red, green and blue values fall within the range of 0 to 1.(9)

Transformation between the RGB and HSV models is quite straightforward. The three-dimensional representation of the HSV model involves a geometrical transformation of the RGB cube. If one views the cube along the diagonal from the white vertex to the black vertex, the cube's outline forms a hexagonal shape. The boundary of the hexagon represents the hues (H), which are expressed as angles about the vertical axis. The main diagonal of RGB space becomes the value (V) axis in HSV space. Saturation (S) varies as one moves from the center of the hexagon to its boundary.

According to Bertin (4), the effectiveness of different visual variables depends on data type, whether nominal, ordinal, interval or ratio. Through the selection of appropriate visual variables based on data type, the cartographer attempts to devise a symbology that effectively communicates spatial data to the map user.(15) Color is often used in cartography to portray attributes of kind or quality. Hue is the color dimension most logically used for such nominal data. For quantitative data, including indices of data quality, cartographic conventions are less well-defined. Hues with the same value are inappropriate for such data, as they are not visually ordered.(4) However, either saturation and value may be used for such data, as either of these variables can be used to create a visually-ordered scale.(13) Humans tend to equate higher saturation with greater magnitude.(15) Greater magnitude is also suggested by lower value, although this tendency is influenced by the background color.(1)

Variations in saturation can be used for visualization of data quality.(2) Brewer (5) devised such a method based on a reduction in saturation in areas of low quality. Schweizer and Goodchild (16) created a bivariate choropleth map in which variations in value and saturation were used together to show variations in data quality and the magnitude of a thematic variable. Their research indicated that although humans can distinguish between variations in hue, saturation and value, they do not necessarily equate the amounts of each these three components with the magnitude of the mapped variables. Another problem is that the maximum number of different values that can be used effectively on a map (approximately 7) is somewhat lower than the number of hues that can be used (approximately 15).(13)

3.2. AUDITORY VARIABLES

Sound is a vibratory disturbance in the pressure and density of a fluid, or in the elastic strain in a solid, capable of being detected by the organs of hearing. Such vibratory disturbances can be viewed as a series of sine waves whose properties define the character of the sound. One property of sound is termed frequency and is defined as the number of completed cycles of vibrations that occur in a one-second interval. The unit of measurement of frequency is the Hertz. For example, if 1000 vibrations occur per second, the frequency of the sound would be expressed as 1000 Hertz (Hz) or 1 kiloHertz (kHz). Human perception of frequency is referred to as pitch. In general, the higher the actual frequency, the higher the pitch that is perceived.

Another property of sound is amplitude, which refers to the distance that a vibrating medium travels during a vibration. Human perception of amplitude is termed loudness.(7) In general, the greater the amplitude, the louder the sound is perceived to be. The physical measure of human perception of loudness is called intensity. Intensity is measured in sound pressure units which provide a measurement of the force of a sound vibration over a given unit area. In physics, dynes per square centimeter is the unit of measure. In order to simplify the reporting of intensity values, a logarithmic scale is used using a unit called the decibel (dB).

One aspect of sound that relates to both perceived loudness and to perceived tone is duration. If duration is less than about 200 milliseconds then intensity must be increased for the sound to be perceived. Similarly, perception of pitch is affected by duration. In order to discern a sound's pitch, a tone must be generated for a minimum of 0.5 to 1.0 seconds. Otherwise the perceived pitch only registers as a 'click'.(8)

The human ear is responsive to a wide range of frequencies, varying from 20 Hz to over 20,000 Hz. Human hearing is most sensitive in the ranges of 1000 to 3000 Hz and 40 to 50 dB. As frequencies drop below 500 Hz, hearing sensitivity decreases. Between 3,000 and 12,000 Hz sensitivity decreases gradually. Sensitivity shows a sharp drop above 12,000 Hz.(8)

Little work has been done on the use of sound in cartography, although a few researchers have suggested that it could be incorporated quite easily into computer mapping systems.(14) Likewise it has been suggested that changes in pitch might be used to portray variations in data quality.(2)

4. SYSTEM IMPLEMENTATION

This section describes the design and operation of the data quality visualization system. The system uses both visual and auditory variables to allow the user to examine the data quality component of a digital soil map. The system is implemented under a graphical windows environment that provides tools for manipulating maps displayed in different windows on the computer screen.

The system displays four windows simultaneously:

(1) a map window in which the soil map is displayed and mapping units are shaded in different colors;

(2) a data quality window in which the boundaries between mapping units are displayed, but no shading is used;

(3) a report window in which various statistical parameters are displayed; and

(4) a legend window displaying the colors and associated legend categories for all mapping units displayed in the map window.

Underlying the map is a grid-based data quality model. Each cell in the grid is associated with a quantity expressing the data quality at that location. The user is able to explore data quality over the spatial domain of the database with the aid of a pointing device such as a mouse. The path traversed by the mouse is recorded in the data quality window by plotting the locations of the mouse as it is moved over the surface of the map window. The color of the symbol plotted at each location reflects the data quality value corresponding to the grid cell at the current mouse location. To represent different levels of data quality, a visually-ordered scale is produced by varying the saturation and value components, while holding hue constant. A double-ended scale is used for quantities that deviate on either side of zero (e.g., positive and negative deviations).

The system is also capable of emitting an auditory signal when a mouse button is pressed. This signal supplements the visual representation of data quality in the data quality window. The duration of the signal (frequency of clicks per second) is inversely proportional to the quality of the data at the current mouse location. This approach is based on a "geiger counter" metaphor.

The report window displays the visually-ordered data quality scale with a marker that moves to identify the value of the current mouse location. A profile is also drawn to show variations in data quality over the path traversed by the mouse. The profile is dynamically updated as the mouse moves.

Data quality in the context of soil maps is a multifaceted problem. Multiple dimensions of data quality exist, and it seems reasonable to assume that different users may be interested in only a subset of these quality dimensions as a function of their objectives in using soil maps. The system is therefore flexible enough to allow different data quality models to be employed. For illustrative purposes, the model presented in the example below is based on apparent errors in published slope values for each mapping unit. (Slope is one of the criteria used to delineate mapping units.) A Soil Conservation Service soil map of Centre County, Pennsylvania, is used in the example.

Figure 1 shows the procedure used to derive the grid-based data quality model. The cell values in the model indicate the deviations between actual slope values derived from a digital elevation model and the published slope values for each soil mapping unit. The published values define the acceptable range of slope values within a given mapping unit. Negative deviations indicate that the actual slope value for the cell is lower than the acceptable lower limit. Positive deviations indicate that the actual slope value is higher than the acceptable upper limit.

Figure 2 is a screen dump showing the configuration of the system. The screen dump depicts the map, data quality and report windows. The legend window is hidden. The screen dump is unable to display the different colors seen on the computer screen. Likewise it is unable to portray the dynamic nature of the display and the way in which the user interacts with the system. Several circles of various tones can be seen in the data quality window. These circles are plotted dynamically as the mouse is moved across the surface of the map window. On the computer screen, the tone of these circles is defined in terms of the data quality value in the grid cell at the current mouse location. As shown in the report window, a double-ended scale is used to indicate negative and positive slope deviations. A red hue is used for positive deviations and a blue hue for negative deviations. The report window also shows a profile of data quality values for the path traversed by the mouse. The profile is updated automatically as the mouse is moved.

5. CONCLUSION

The main features of the system that facilitate data quality visualization are summarized below.

(1) The system is dynamic. It makes use of animation to facilitate exploratory analysis of data quality over space. This is in contrast to approaches based on the production of a single hard-copy map to portray data quality variations. The exploratory paradigm allows the user to have direct control over those aspects of data quality that are of interest, and facilitates a greater understanding of the nature of data quality variation as it allows the user to test informal hypotheses.

(2) The system separates map data from information about data quality. Attempts to create bivariate maps portraying both map data and information about data quality have proven to be difficult for users to interpret, or are restricted to a small number of classes of data quality.(5,16) The system avoids these problems by displaying information about data quality in a separate window from the map data themselves.

(3) The system relies on both visual and auditory variables to portray variations in data quality. This complementarity enhances the ability to communicate information about data quality.

(4) Various models of data quality variation can be employed. The system is not restricted to one data quality model, as in the case of hard-copy maps. The user is free to explore different models to gain a better appreciation for differences in data quality as depicted by these models.

Future work will involve testing of the effectiveness of the system as a means of communicating soil map quality information. Other data quality models will also be devised and tested, including models based on soil type inclusions and polygon purity, models based on the classification error matrix and other thematically-disaggregated models, and models based on mapping unit boundary width.

6. ACKNOWLEDGMENTS

This work was carried out in conjunction with the "Visualization of Data Quality Challenge" sponsored by the National Center for Geographic Information and Analysis. Data were obtained from the National Soil Survey Center, USDA-SCS, Lincoln, Nebraska.

7. REFERENCES

1. Antes, J.R. & Chang, K. 1990. An empirical analysis of the design principles for quantitative and qualitative area symbols. Cartography and Geographic Information Systems 17(4): 271-277.

2. Beard, K., Buttenfield, B. & Clapham, S. (Eds.) 1991. Visualization of Spatial Data Quality. NCGIA Technical Paper 91-26. Orono, ME: National Center for Geographic Information and Analysis.

3. Beatty, J.C. & Booth, K.S. 1982. Tutorial: Computer Graphics. Silver Spring, MD: IEEE Computer Society Press.

4. Bertin, J. 1983. Semiology of Graphics. Madison, WI: University of Wisconsin Press.

5. Brewer, C. 1992. Color selection for geographic data analysis and visualization. Presented at GIS/LIS '92.

6. Burrough, P.A. & Bie, S.W. (Eds.). 1984. Soil Information Systems Technology: Proceedings of the 6th Meeting of the ISSS Working Group on Soil Information Systems.

7. Deutsch, L.J. & Richards, A.M. 1979. Elementary Hearing Science. Baltimore, MD: University Park Press.

8. Everest, F.A. 1989. The Master Handbook of Acoustics. Blue Ridge Summit, PA: TAB Books.

9. Foley, J.D. & Van Dam, A. 1982. Fundamentals of Interactive Computer Graphics. Reading, MA: Addison-Wesley.

10. Goodchild, M.F. & Gopal, S. (Eds.) 1989. Accuracy of Spatial Databases. Basingstoke: Taylor and Francis.

11. Hearn, D. & Baker, M.P. 1986. Computer Graphics. Englewood Cliffs, NJ: Prentice-Hall.

12. Lewell, J. 1985. A-Z Guide to Computer Graphics. New York, NY: McGraw-Hill.

13. Mersey, J.E. 1990. Color and thematic map design: The role of color scheme and map complexity in choropleth map communication. Cartographica, 27(3).

14. Ormeling, F. 1993. Ariadne's thread -- structure in multimedia atlases. Proceedings, 16th International Cartographic Conference, 1093-1100.

15. Robinson, A.H. 1967. Psychological aspects of color in cartography. International Yearbook of Cartography, 7, pp. 50-59.

16. Schweizer, D. & Goodchild, M.F. 1992. Data quality and choropleth maps: An experiment with the use of color. GIS/LIS 92, 686-699.

17. Western, S. 1978. Soil Survey Contracts and Quality Control. Oxford: Oxford University Press.

18. van der Wel, F. 1993. Visualization of quality informations as an indispensible part of optimal information extraction from a GIS. Proceedings, 16th International Cartographic Conference, 881-897.

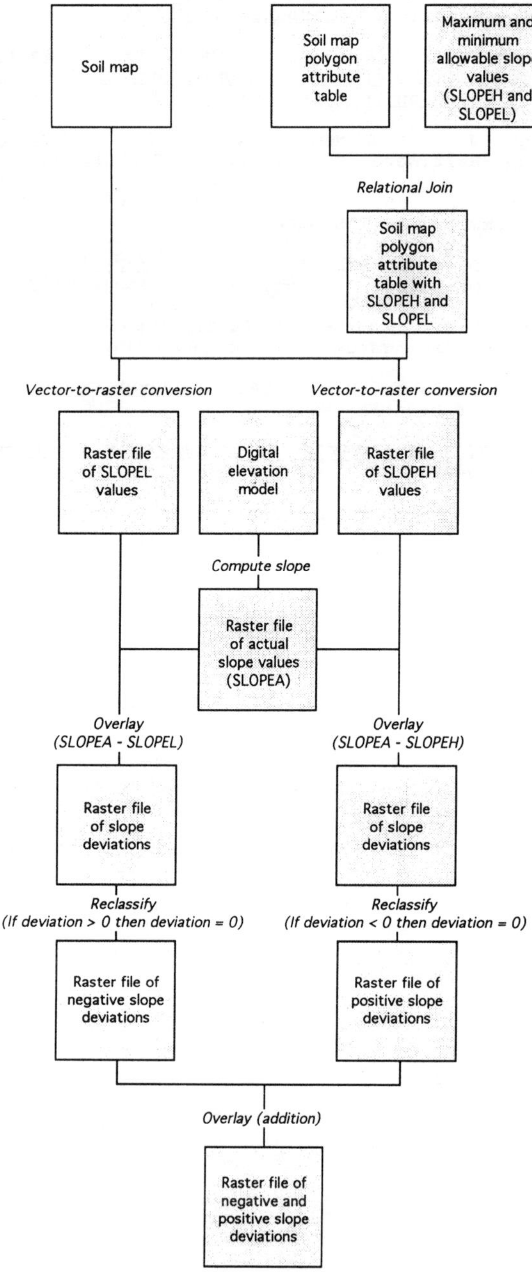

Fig. 1. Data Flow Used to Derive Data Quality Model for Slope

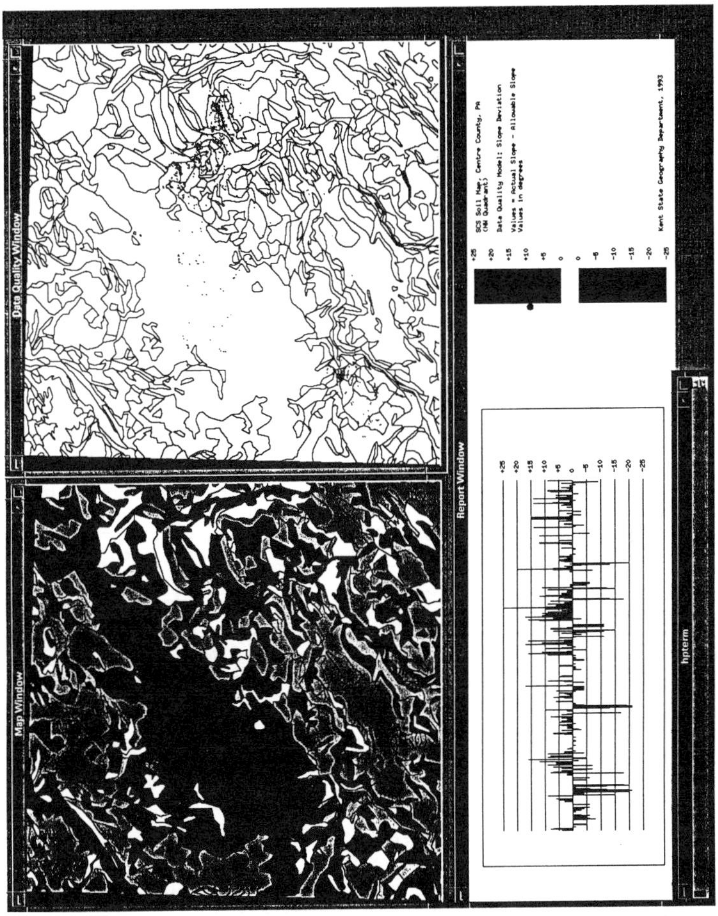

Fig. 2. Screen Dump Showing System Configuration

SCS WETLAND RECERTIFICATION USING SPOT DATA

Susan M. Berta
Indiana State University
Terre Haute, Indiana 47809

Timothy A. Gress
ITD/Space Remote Sensing Center
Building 1103, Suite 118
Stennis Space Center, MS 39529

1. INTRODUCTION

The alarming loss of wetland habitats due to agricultural encroachment has resulted in government legislation mandating the monitoring of wetlands as set forth in the Food and Security Act (FSA) of 1985 (Public Law 99-198, Title 12). Natural wetlands and areas which experience 50 percent frequency of flooding over 15 consecutive days during the growing season can not be converted to cropland after December 23, 1985 without loss of support according to FSA provisions.

The 1990 Farm Bill calls for the U.S. Department of Agriculture - Soil Conservation Service (SCS) to conduct periodic reviews and updates of existing wetland inventories. Every year National Headquarters selects 300 counties for State Offices to do a quality review and to recertify a statistical sampling of enough wetland determinations to ensure 95 percent reliability in existing wetland inventories. These reviews identify suspect wetland delineations and alert National Headquarters to potential conversions of wetlands to agricultural lands. The intent of this research is to assess the potential for identifying a consistent procedure by which SPOT data is utilized as baseline input for these reviews.

The following paper describes a wetland recertification and sampling pilot project completed for the SCS by the Indiana State University Remote Sensing Laboratory (ISURSL) in conjunction with ITD/Space Remote Sensing Center. Classification results utilizing SPOT digital data were compared with an existing SCS wetland inventory for a study area located in Delaware County, IN.

2. BACKGROUND

Northeastern Indiana experienced the full impact of Pleistocene glaciation resulting in a modern day landscape where well-integrated drainage is lacking and wetlands abound. By the mid-1980's Indiana had drained from 85 to 90 percent of their wetlands to create cropland and pasture. Young and flat glacial terrain with soils dominated by heavy silt loams in an area with an average 35 inches of annual precipitation results in a high percentage of soils aquic in character. Delaware County is located in northeastern Indiana and lends itself to an investigation focussing on, in part, identifying hydric soils related to wetland areas.

The ability to accurately identify water features and changes in water features is, in part, dependent upon the meteorological/precipitation characteristics prior to data acquisition. Approximately three inches of precipitation fell during the fifteen day period prior to the October, 1986 date of acquisition. The July, 1987 and May, 1988 dates received less rainfall (1.74 and 0.69 inches, respectively). The summer of 1988 has been characterized as a period of "drought" yet, note that the 1988 image was acquired in May; prior to the onset of dry conditions throughout this region. The June, 1989 date received a total of 3.90 inches over the fifteen day prior time period with 1.01 inches recorded in the four days prior to data acquisition.

Given the disparity in precipitation totals among the dates provided for this investigation, the following generalizations warrant consideration: October, 1986 and June, 1989 dates may tend to over-represent water features on the landscape (with the June date having the highest errors of commission) while the July, 1987 and May, 1988 dates may be more consistent in identifying perennial water features (fewer errors of commission) while the October and June dates may be more suited to finding temporary/standing water features.

3. DATA ANALYSIS

Multitemporal SPOT data were used to assess the changes in spatial patterns of woody vegetation and standing water in northwestern Delaware County. Data supportive of the SPOT scenes (10/7/86, 7/20/87, 5/1/88, and 6/6/89) included 1:12,000 orthophoto maps and 1:24,000 topographic maps. The Delaware County Soil Survey (2) was available as was a digital copy of the SCS Wetlands classification (compiled from the interpretation of aerial photography) and TIGER DLG files.

Each individual date of satellite imagery was analyzed using ERDAS (1) or MultiSPEC (4) software. High dimensional clustering (200 cluster classes) was used in an unsupervised classification approach. The spectral statistics were then input into a maximum likelihood classification algorithm. Viewing the classification results and visually integrating other information such as spectral curves, orthophoto maps, and topographic maps, as well as incorporating information gathered from a field visitation, enabled the identification of individual land cover classes. The land cover types included open water, wet land/wet soil, and forest/woody vegetation. Where suitable, classified data were smoothed using a 3x3 or 4x4 filter to reduce data complexity.

The classified data were then geocorrected to the UTM coordinate system. The final four classified scenes were incorporated as GIS files in ERDAS and were overlaid and processed to develop additional classified data sets indicating multitemporal (1986-1989) changes involving potential forest-to-agricultural land and wetland-to-agricultural land cover changes. Road network information was overlaid using TIGER DLG files to facilitate future field verification of land cover changes.

Field verification of the water classifications was not possible given the temporary nature of the standing water features. Accuracy assessments are presented using qualitative discussion and the polygon intersection method described in Gress, et al. (3). The polygon intersection method requires the wetland classification and the SCS inventory be converted from raster forms to ARC/INFO vector coverage. The two classifications were converted to coverage and the number of classification polygons intersecting the SCS inventory polygons were counted using the frequency program. This method of accuracy assessment has no true means of identifying commission and omission errors. Qualitative discussions of these errors based on visual comparisons of the classifications are included in the assessment of accuracy.

4. RESULTS AND CONCLUSIONS

The SCS Wetland Inventory includes five classes related to wetlands in this area. The ISURSL digital data classifications are not intended to mimic or replicate the SCS Wetland Inventory for this area but rather to identify deforested sites, drained sites, and previously excluded wetland sites.

Forest/woody vegetation was initially mapped to identify all major areas of forest which were dense and homogeneous. Sparse forests (25 - 75% crown cover) and forests distributed in narrow tracks along watercourses, ditches and windbreaks were less well identified. The accuracy in identifying forest pixels as shown on orthophoto ground truth was approximately 95 percent for each date.

The multitemporal results derived from a GIS overlay of the 1986 and 1989 forest classifications, using a 4x4 filter, indicated that 5.0 percent of the area's forests were potential sites where forest had been cut and converted to cropland by 1989. These areas warrant field verification to assess FSA compliance. When using a 3x3 filter, the potential forest-to-agricultural land cover change increased to 7.2 percent while the unfiltered classification results yielded an approximate 11 percent change from forest to agriculture. Classification errors associated with forest cover included an estimated 16.7 percent of the area classified as forest in 1989 did not exist as forest in one or more of the previous dates. Consistently 78.2 percent of the forests were coincident throughout the four date study.

Water features identified throughout the study area refer to permanent as well as temporary or shallow water features oftentimes having high suspended sediment concentrations. The temporal approach provides data on areas known to have surface water present in 1986 and any possible water cover changes that may have occurred up to June 6, 1989. In general, the October, 1986 and May, 1988 classifications provided excellent accuracy in identifying open water features (>95 percent accuracy) while the July, 1987 classification was less effective (78 percent accuracy). The June, 1989 classified data is representative of the time period with the greatest amount of standing water present. All known open water sites were identified as well as many locations associated with the "Wetland" designation on the SCS inventory. Generally, the additional standing water features were not limited to Wetland designated sites: many of the larger contiguous areas co-registered with ditches, streams, edges of fields, and several agricultural fields.

A composite map was developed to identify water found on a minimum of two of the individual classifications using the first three study dates of data. Approximately 95 per cent of the artificial wetlands identified in the SCS inventory are mapped as water features using this composite technique. The few unaccounted for "water" pixels are associated with SCS designated wetlands. Several locations were identified as having standing water not identified in the SCS inventory and warrant field investigation for reconsideration as SCS wetland sites. A pond was consistently identified throughout the study period in the ISURSL classification (as well as on the orthophoto quadrangle) yet is not identified in the SCS Inventory. Several patches of standing water are found in agricultural fields in the three date composite of water features that are not identified on the SCS Inventory that warrant investigation for inclusion as SCS classes "Wetland" or "Artificial Wetland".

The water features noted in the classifications for the first three study dates were next compared to water features mapped in June, 1989. This comparison allowed the identification of locations which were: (1) consistently identified as open water (SCS artificial wetland), (2) potentially drained by June, 1989 (non-wetland areas), and (3) potential wetland sites which were overlooked in the previous wetland inventory. Given the conservative approach used to create this composite (minimum two date occurrence of water) and the "wet" June date, this procedure provides great potential for identifying suspected drained wetlands warranting immediate field verification.

The final assessment in this investigation involved comparing the SCS Wetland Inventory with the ISURSL forest and water classifications. The ISURSL base map consisted of FOREST as identified on the October, 1986 classification and the three date composite of water features where water appearing two or more times over this three year period was included. A pixel location identified as forest and water in the individual classifications was assigned as forest in this composite data set. An accuracy assessment noting the frequency of polygon intersection when comparing the ISURSL composite classification to the SCS Inventory reveals a 77.9 percent overlap: 77.9 percent of the SCS Inventory polygons intersected either forest or water pixels identified in the ISURSL classification.

Utilizing digitally classified SPOT data to consistently identify forest/woody vegetation and changes in this land cover proved to be effective with an estimated accuracy greater than 78 percent. Open water features were accurately identified using the three date composite data set (>95 percent accuracy) establishing SPOT data as a viable data source for consideration in future SCS wetland inventory reviews. Although of somewhat degraded resolution, the use of Landsat TM data warrants investigation to improve the mapping accuracy of forest/woody vegetation. Landsat TM, with mid-infrared band width data, may aid in reducing the problem of classifying transitional forests, particularly during June through August when dense wet cropland was, can be confused with some forest types. The sensitivity of the two mid-IR bands may provide additional aid in identifying water-saturated agricultural fields.

Using digital classification of SPOT data to identify areas for possible updating to wetland and/or artificial wetland classes appears feasible (pending field verification). Access to a date later in the calendar year (September date) may be even more appropriate for this task. The potential for identifying areas which have changed from wetlands to non-wetlands appears feasible (pending field verification) with a recommendation for future studies to incorporate a "wet" date as the latest date in the multitemporal sequence.

Overall, the use of multitemporal SPOT data to identify forests, open water, and changes in these cover types provides an accurate means of developing maps for the periodic review and updating of existing SCS wetland inventory data (pending field verification). The SCS Wetland Inventory classes of Converted Wetlands, Farmed Wetlands, Prior Converted Wetlands, Minimum Effects, and Non-Wetlands could not be distinguished per se using digital image processing of SPOT data but relevant information concerning changes in forest cover and open water/saturated soils provides a basis by which to indirectly and accurately assess changes in wetlands in this area.

5. REFERENCES

1. ERDAS, Inc. 1991. PC-ERDAS Software, Version 7.5. Atlanta, GA: ERDAS, Inc.

2. Gress, T.A., D. Kettler, C. Mao, L. Wright, and S. Berta. "Wetland Recertification and Sampling Using Satellite Remote Sensing." Proceedings Volume from the World Space Congress. Washington, D.C. (in press).

3. Huffman, K.K. 1972. "Soil Survey of Delaware County, Indiana." United States Department of Agriculture. Washington, D.C.: Superintendent of Documents. U.S. Government Printing Office.

4. Landgrebe, D. and L. Biehl. 1992. "An Introduction to MultiSpec." Version 6.30.92. West Lafayette, IN: Purdue Research Foundation.

MODELING THE EXPANSION OF SPARTINA IN THE PALIX ESTUARY

John A. Harrington, Jr.
Indiana State University
Terre Haute, Indiana 47809

Lisa M.B. Harrington
Eastern Illinois University
Charleston, Illinois 61920

1. INTRODUCTION

Species introductions are known to produce significant impacts on the new environment as a niche is established. (5) Willapa Bay in Pacific County, southwestern Washington, provides an example of an intentional and beneficial introduction of the eastern oyster (Crassostrea virginica), which was accompanied by the unintentional introduction of cordgrass (Spartina alterniflora). (6) Changes in the reproductive status of Spartina during the last 15-20 years have made Spartina a potential ecosystem-altering species. (11)

Willapa Bay is a large estuary and maintaining the quality of the waters and the substrate are a major concern for the county, because the estuary provides a critical natural resource base. Willapa Bay, which is known globally for the production of oysters, has been described as "undeniably the cleanest estuary still remaining on the entire U.S. coast." (1) Twenty river systems drain into the bay (10); many of these are short, but carry a high annual flow in a region where annual precipitation totals range from 165 to 254 centimeters. (7) The drainage area for the estuary is about 2850 square kilometers, with low levels of agricultural and urban land use (less than 5 percent of the drainage area, combined). (9) The primary land use for the watershed is forestry. A low population density contributes to the relatively high environmental quality of the bay.

Willapa Bay continually changes appearance with the tidal cycle. The ebb and flow of tides and the slow washing of nutrients into the bay from the Willapa Hills helped create a highly productive estuarine environment. At high tide, Willapa Bay is a wide expanse of open water with forests, and farmland on the former high marsh environments, occupying its margins. The vast extent of open water at high tide conceals the shallow nature of the bay. At low tide, more than 50 percent of the bay substrate is exposed (7) as seemingly barren mudflats. The mudflats have historically been highly productive for a number of saltwater species. A major oyster industry existed here in the latter 1800s, when oystermen harvested native oysters (Ostrea lurida) to supply the San Francisco market. Oysters continue to be a very important product.

2. SPARTINA EXPANSION

The cordgrass that was introduced into Willapa Bay, Spartina alterniflora, is native to estuaries of the east coast of the United States. Following its introduction in 1894, the initial spread of Spartina was very slow. It apparently caused few problems and received little attention. In 1942, the Willapa National Wildlife Refuge manager noted that local oystermen had expressed concern about a potential Spartina problem and in 1979 Washington Department of Wildlife field biologists recommended eradicating Spartina from Willapa Bay. (8) However, a 1981 U.S. Fish and Wildlife Service publication that attempted to point out important research questions for Willapa Bay did not list the expansion of Spartina among "the more significant questions about the environment of Willapa Bay," although there was an awareness of a number of introduced species, including Spartina, in the estuary. (7)

Spartina has undergone a rapid exponential expansion (2,6) that is characteristic of biological populations with little competition. Prior to 1980, seedlings were infrequently seen in Willapa Bay and the species maintained itself primarily through vegetative reproduction. (4) The 1982 areal coverage estimate was 800 acres (Figure 1). By 1989, it was reported that over 2400 acres of Willapa Bay were affected by Spartina; potential habitat is over 50,000 acres. (2) Spartina alterniflora in Willapa Bay is now producing viable seed. (11) Uncertainties exist regarding the exact reason(s) for the change in reproductive status, but recent warmer temperatures may be a contributing factor. (12) Spartina flowers in late summer, and the decade of the 1980's was an extended period of above normal July and August temperatures (Figure 2). Decreased water depths associated with sediment deposition has also been suggested as a contributing factor in increasing water and substrate temperatures. (12)

3. MODELING SPARTINA EXPANSION

Modeling change in ecosystems requires knowledge of several important aspects of the functioning of the system. Factors that need to be considered include: knowledge of the reproductive rates of the species involved, knowledge of the habitat requirements for individual species, and information concerning the competition between species that may affect the geographic pattern of species expansion/contraction. Modeling the complex dynamics of Spartina alterniflora expansion in Willapa Bay can be approximated using several seemingly realistic assumptions about the ecosystem. First, the available niche or potential habitat for the expansion of Spartina must be defined. In Willapa Bay, the subtidal wetlands (areas exposed at low tide) are thought to be available for colonization. (8) Since these areas are currently sparsely vegetated or not vegetated, the expansion can be considered similar to primary vegetative succession onto newly created habitat. Because these estuarine subtidal wetland areas have little to no natural vegetation, Spartina has few, if any, competitors for the available habitat. In determining the location for new colonies of Spartina on the Willapa Bay mudflats, a random model of species dispersal can be assumed, because wind and wave action are capable of moving viable seed from existing sources throughout the system. The outward expansion of existing colonies of Spartina needs to be accounted for and has been measured at .5m/yr. (11) One scenario for the potential changes that might occur to the Willapa Bay mudflats with Spartina expansion can be created by extrapolating from the existing exponential growth trend (Figure 1). Estimates of the acreage of Spartina for 1992, 1996, 2000, and 2004 were generated using the standard exponential growth curve for biological populations. (2)

For some individuals, particularly visual learners, a map of "how things might be," rather than graphs, tables or statistics, can provide a powerful communication message. A primary concern in this modeling effort was to generate maps for one (scientifically sound) scenario of the future. Geographic information systems provide a useful platform for producing visually-oriented data products. Polygon files marking the limits of wetland types in 1982 for the appropriate USGS topographic map quadrangles were obtained by acquiring the US Fish and Wildlife Service National Wetlands Inventory (3) in ARC/INFO format from the Washington Department of Ecology. The data for the ten quadrangles that encompass Willapa Bay were joined in ARC/INFO and then converted into raster or grid cell format. Working in the ERDAS raster GIS environment, the 172 different categories of wetlands were generalized or reclassified into three major categories needed for the modeling effort: 1) existing aquatic macrophytes (Spartina colonies and othe tall aquatic plants), 2) available habitat (subtidal wetlands), and 3) all other areas (including deep water and dry land). At this point, it was decided to work with a small area within Willapa Bay. A subset that includes the Palix Estuary portion of Willapa Bay was selected (Figure 3a). The amount of subtidal wetland in the Palix Estuary subset (compared to the whole of Willapa Bay) was determined and used to assign a proportional acreage of the exponential growth at the various time steps in the scenario. Next, a GIS function that replaces a specified number of random grid cells of an existing category with a new category was used to produce a cartographic scenario for three future years 1996, 2000, and 2004 (Figures 3b-d). For the time step between 2000 and 2004, the outward expansion of existing colonies was incorporated. A portion of the growth (new grid cells) for the 2000 to 2004 time increment was assigned to grid cells that were adjacent to areas of Spartina as produced in the modeling scenario for 1989. The modeling effort does not take into account any human efforts to alter the system, such as mechanical harvesting, chemical control, or biological control. (8)

The four maps that comprise Figure 3 provide a visual indication of one scenario for rapid on-going change in a portion of Willapa Bay. The changes that are occurring in Willapa Bay are well known in the local region, the state capitol, and by scientists at the national level. As with any issue concerning the environment and change, there are a variety of opinions regarding what should be done or not done to arrest the expansion of Spartina in Willapa Bay and other estuaries of the Pacific Northwest. One scientific publication (8) indicates that the pace of change is so rapid and that Spartina expansion has advanced to such a point in Willapa Bay that we may end up "living with it." It is hoped that the cartographic products produced through this modeling scenario will provide some insights regarding one possible future outcome.

4. REFERENCES

1. Atkinson, J. (Willapa Wildlife Refuge), 1991. Statement made in "Willapa Bay: A Delicate Balance." Portland, OR: Odyssey Productions, Inc. (video)

2. Colby, M.E., 1992. Eco-Accounting for Willapa. Published by the Willapa Alliance, Ocean Park, WA 98640

3. Cowardin, L.M., V. Carter, F.C. Golet, and E.T. LaRoe, 1979. Classification of wetlands and deepwater habitats of the United States. USFW FWS/OBS-79/31.

4. Frenkel, R.E., and L.M. Kunze, 1984. Introduction and spread of three Spartina species in the Pacific Northwest. Abstracts, Assoc. of American Geographers Annual Meeting, Washington, D.C.

5. Goudie, A., 1990. The Human Impact on the Natural Environment. MIT Press, Cambridge, MA 388p.

6. Harrington, L.M.B., and J.A. Harrington, Jr., (1992). "Willapa Bay, Washington: Environmental Concerns and Natural Resource Management." Pages 162-169 In: Geographic Perspectives on the Social and Economic Restructuring of Rural Areas, M.D. Nellis (ed.), Kansas State University, Manhattan, KS.

7. Hedgpeth, J.W., and S. Obrebski, 1981. Willapa Bay: A Historical Perspective and a Rationale for Research. Washington, DC: Office of Biological Services, US Fish and Wildlife Service.

8. Mumford, T.F., P. Peyton, J.R. Sayce, and S. Harbell (eds.), 1990. Spartina Workshop Record. Published by the Washington Sea Grant Program, University of Washington, Seattle, WA 73p.

9. National Oceanic and Atmospheric Administration, 1990. Estuaries of the United States: Vital Statistics of a National Resource Base. Strategic Assessment Branch, Ocean Assessments Division, Office of Oceanography and Marine Assessment, National Ocean Service, Rockville, MD: National Oceanic and Atmospheric Administration.

10. Odyssey Productions, Inc., 1991. Willapa Bay: A Delicate Balance. (1991), Portland, OR (video)

11. Sayce, K., 1988. Introduced cordgrass *Spartina alterniflora* Loisel. in saltmarshes and of Willapa Bay, Washington. USFWS FWSI-87058 TS, 70p.

12. Sundberg, S., 1992. Spartina: Distribution, Biology, and Ecology. Final Report Element A, Submitted to Washington State Department of Ecology, Olympia, WA 30p.

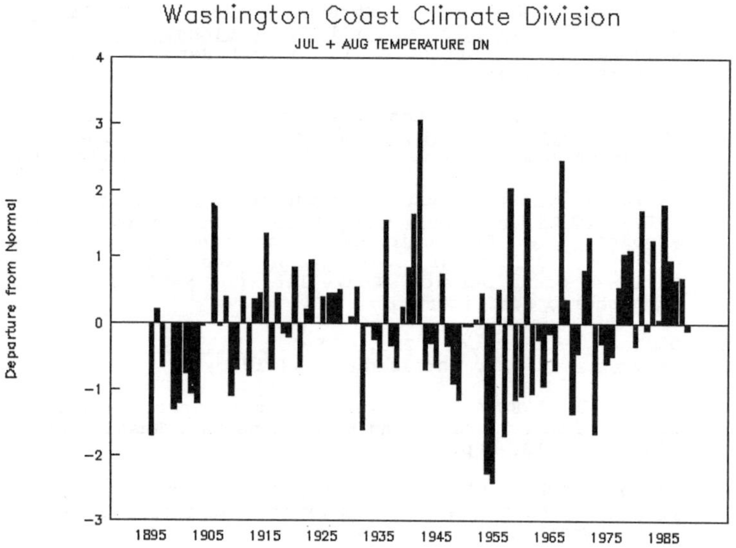

FIGURE 1. Time series of July plus August temperature deviations (°F) from the long term average.

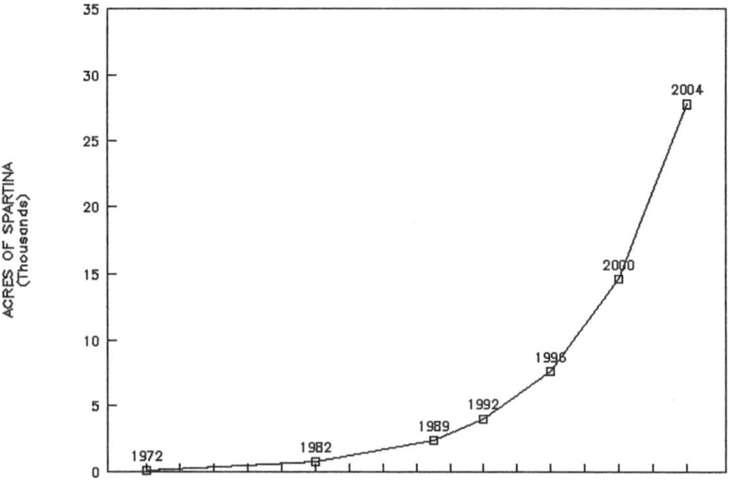

FIGURE 2. Measured (1982, 1989) and modeled (1992, 1996, 2000, 2004) changes in the acreage of *Spartina* in Willapa Bay, Washington.

A. 1982

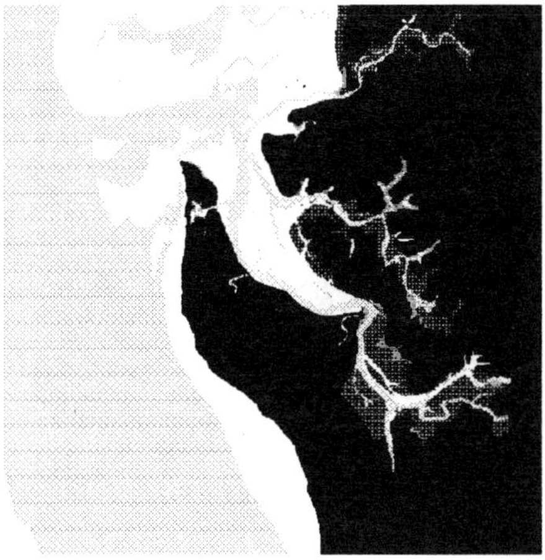

B. 1996

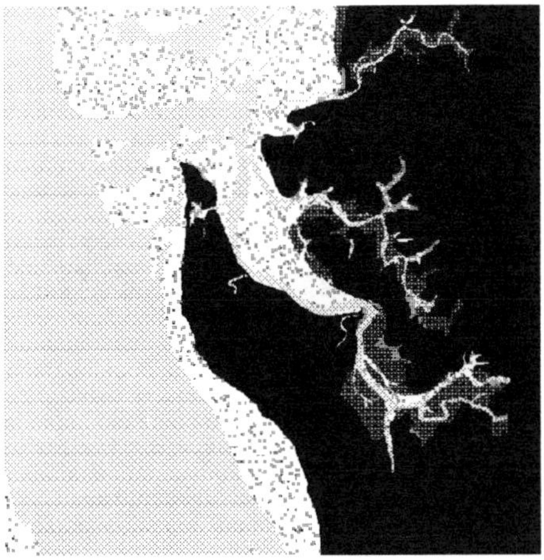

FIGURE 3. The distribution of *Spartina* in the Palix Estuary portion of Willapa Bay for 1982 and 1996 based on an exponential growth model (see text). Black areas are upland, dark gray represents existing or modeled areas of aquatic macrophytes (*Spartina*), light gray areas are deep water portions of Willapa Bay, and white areas are mudflats.

C. 2000

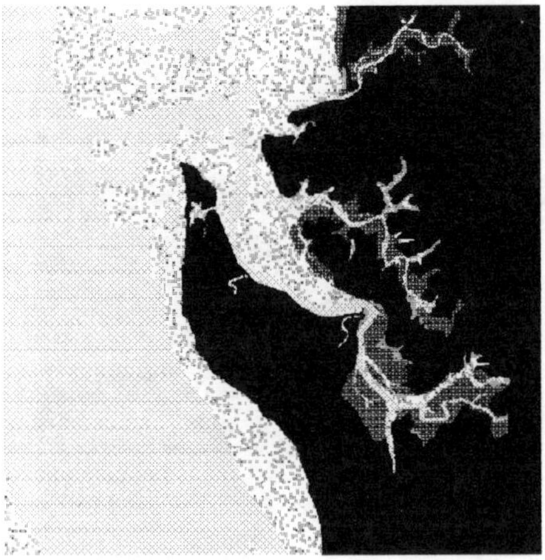

D. 2004

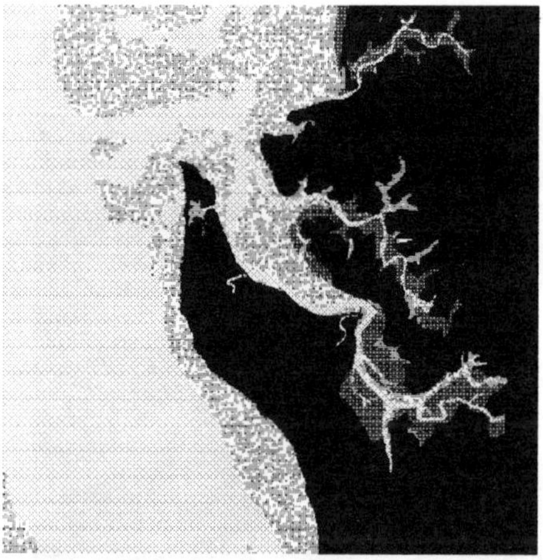

FIGURE 3. The distribution of *Spartina* in the Palix Estuary portion of Willapa Bay for 2000 and 2004 based on an exponential growth model (see text). Black areas are upland, dark gray represents existing or modeled areas of aquatic macrophytes (*Spartina*), light gray areas are deep water portions of Willapa Bay, and white areas are remaining areas of mudflats.

THE NORTH AMERICAN FREE TRADE AGREEMENT AND ITS IMPACT ON THE REGIONAL DEVELOPMENT OF THE CENTRAL RIO GRANDE VALLEY

Robert J. Czerniak
New Mexico State University
Las Cruces, New Mexico 88003

1. INTRODUCTION

El Paso, Texas has been a consistent focus of international trade and migration between the United States and Mexico. The signing of the North American Free Trade Agreement (NAFTA) has once again raised expectations that El Paso and the central Rio Grande Valley are on the verge of significant economic base expansion from increased border activity. The paper examines current regional efforts in infrastructure development, urban planning and zoning in light of potential impacts from the NAFTA agreement.

2. THE NORTH AMERICAN FREE TRADE AGREEMENT

The North American Free Trade Agreement (NAFTA) which was signed by the leaders of Mexico and the United States on December 17, 1992 promises to increase trade between the two countries. NAFTA will affect two major issues in international trade--market access and trade rules. A series of new issues relating to the NAFTA have also been identified. They are: investment, intellectual property, environment, labor adjustment and maquiladoras. (8)

The United States has responded to the NAFTA in a variety of ways. There are the newspaper reports and congressional concerns about job loss, environmental degradation and increased illegal immigration. (15) In an attempt to respond rationally to the potential effects from the agreement, both federal and state governments have developed plans that will mitigate some of the negative effects. For example, the US Congress passed Section 6015 of the 1992 Intermodal Surface Transportation Efficiency Act which requires that the Federal Highway Administration undertake a study to analyze the preparedness of US border crossings to handle an increase in commercial traffic from NAFTA. (5) The state of Texas alone estimates that a little less than two billion dollars will be needed for highway and bridge improvement as a result of increased traffic from the NAFTA (16).

3. BORDER REGION IMPACT

The region that is anticipated to experience the most direct impacts from the NAFTA is the US-Mexico border extending from Brownsville, Texas to San Diego, California. The largest population concentration along the international border is located within the El Paso, Texas-Juarez, Mexico agglomeration (Figure 1).

The region includes three population centers: El Paso County-515,342, Juarez Municipio-1.2 million and Dona Ana County, New Mexico-135,362. Each center is preparing for potential NAFTA related impacts in the areas of land use planning and infrastructure development. Although it is difficult to separate NAFTA related impacts from impacts caused by other factors, local governments are expediting their planning and development efforts to meet the perceived increased demand. The next three sections discuss how each center will cope with increases in trade, population and economic base.

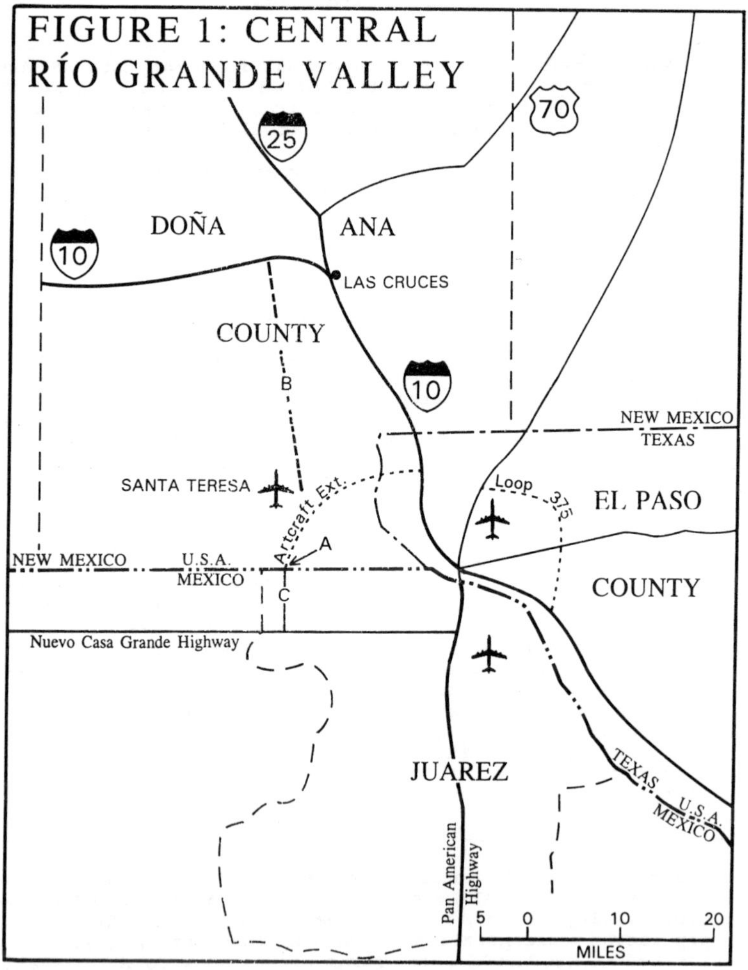

FIGURE 1: CENTRAL RÍO GRANDE VALLEY

4. INFRASTRUCTURE

4.1 TRANSPORTATION-PERSPECTIVE ON THE PROBLEM

NAFTA should make it easier for commercial and passenger traffic to cross at the border. This will impose severe congestion on the existing central Rio Grande Valley transportation system, because it is not adequate to meet current border demand. With the exception of one border crossing (Zaragosa), the El Paso's-Juarez border crossings have operated at or near capacity for many years. Until 1992 Dona Ana County never had a legal border crossing. Although this facility may relieve some of the congestion in El Paso-Juarez, it will not be significant, because the approaches to the new port-of-entry are inadequate on both sides of the border.

4.2 TRANSPORTATION IMPROVEMENTS

4.2.1 El Paso

El Paso has long been a primary crossing point for commercial trade and migration into the United States (11). The major concern of the El Paso business community and city government is to further open its three ports-of-entry to all trade that wish to use them. Twelve projects in El Paso have been identified as being affected by the NAFTA. The total needed to fund all the projects is over $500 million (16). The largest of these is a new loop road around the eastern portion of El Paso. This project alone amounts to a $370 million improvement (hachured line-Figure 1)

4.2.2 Dona Ana County

For the first time Dona Ana County, New Mexico is expecting to benefit directly from increased commercial trade from Mexico. Although Dona Ana County shares a common border with the state of Chihuahua, an official border crossing has never existed with its southern neighbor. During 1992, the Santa Teresa-San Geronimo crossing opened to commercial traffic ("A"-Figure 1).

The short term future of the crossing is limited, because truck access to the facility is poor. Trucks must drive on a series of two lane paved roads through a high income residential area (in Texas) before reaching Interstate 10 (dotted line-Figure 1). The major improvement for transportation in New Mexico is to obtain a multilane paved roadway through Texas which would create new access for Interstate 10. The cost of this project is $9.6 million. (1) For the longer term, an additional $120 million in projects is requested to build a new toll facility ("B"-Figure 1) and upgrade a reliever airport (Santa Teresa-Figure 1).

4.2.3 Ciudad Juarez

Ciudad Juarez may have the most difficulty building transportation infrastructure as a result of impacts from NAFTA. The concentration of population along the border and lack of funding for major investments make it problematic to build infrastructure. The major focus for infrastructure improvement over the past two years has been driven by demands from New Mexico for a paved connection to Dona Ana County's Santa Teresa-San Geronimo border crossing (Figure 1-"A"). Only in 1993 did United States officials receive assurances from the Mexican government that the Mexican access road would be improved so that trucks could drive to the border crossing facility on something more than a dirt road ("C"-Figure 1). Without support from the Mexican central government, it will be impossible for the government of Juarez to keep up with the demands for from Dona Ana County or El Paso.

4.3 OTHER INFRASTRUCTURE-PERSPECTIVE ON THE PROBLEM

Although there are numerous infrastructure issues for all three centers of population related to NAFTA, including new community facilities and solid waste management, the single most important issue is municipal access to ground water, while preserving ground water aquifers. In a region where the annual average precipitation is about seven inches, water is a precious resource. Further complicating the water issue, all three centers draw from essentially the same sources--the Hueco or Mesilla Bolsons. Texas and Dona Ana County have only recently settled a bitter and expensive series of law suits over these aquifers which in the end returned both groups to negotiation. (7) The investment needed to meet water and waste water demands are huge. The best indicator of the need is provided by a series of studies which focused on the infrastructure required for only the low income portions of El Paso and Dona Ana Counties. (12) The total need related to potential impacts from NAFTA and other factors will be much larger than the dollar amounts stated below.

4.3.1 A No Action Alternative

Is it possible for the three population centers to ignore the need to provide potable water and sewer service in areas that are urbanizing and are likely to be affected by NAFTA? According to Laurence Nickey this can only be done as a threat to the health of the region. He indicates that without attention to the water issues, diseases like Hepatitis A, cholera, dysentery, and tuberculosis may become common on both sides of the border. (13)

4.4 WATER AND WASTEWATER IMPROVEMENTS

4.4.1 Texas

In Texas, the estimate to fulfill current needs for urban water and waste water service in low income communities in El Paso County is $134 million. Future growth will require an additional $20 million. (14) The costs do not include the price of purchasing new water rights from farms and ranches in the region.

4.4.2 Dona Ana County

The areas which are expected to experience the most intense development in Dona Ana County are the southern portion of the county centered on Santa Teresa. The state of New Mexico undertook a study similar to the one by the Public Service Board of El Paso to determine the cost of providing municipal water and wastewater service to low income communities in Dona Ana County. Besides Santa Teresa, there were twelve other communities considered in the study. The estimated cost to provide facilities to these areas is $27 million. (12)

4.4.3 Ciudad Juarez

Water and wastewater needs for Juarez are the most difficult to calculate. The magnitude of the problem is serious. Over twenty-five percent of Juarez is not serviced with municipal water. (10) As Laurence Nickey M.D., Public Health Director for El Paso County, noted, "(Juarez) has no sewage treatment facility and an open sewerage ditch parallels the city limits of El Paso for 18 miles". (13) Obviously, the dollar amount needed to meet the demand for the city is overwhelming whether or not NAFTA becomes a reality.

Although no dollar estimates have been made for Juarez, using an adjusted per capita multiplier from the Public Service Board (PSB) study provides a crude indicator of the magnitude of costs for water and wastewater improvements. The PSB study yields a per capita multiplier of $2,680. (14) Let us assume it will cost one-half of the PSB rates for water (PSB=$1200) and waste water treatment (PSB=$1,480) to meet construction costs and Mexican environmental standards. The rates become $600 and $740 respectively. If all of Juarez is to be sewered, the total would be approximately $720,000,000 and the water service would cost $220,000,000. Together, the cost would certainly exceed the estimated one billion dollars.

5. LAND USE-PLANNING AND ZONING

5.1 PERSPECTIVE ON THE PROBLEM

If the NAFTA is successful within the central Rio Grande Valley, it could result in a land use nightmare. One of the central criteria for siting new development is to locate it in an efficient way, trying to coordinate its location with existing and planned infrastructure. In order for this to take place, a comprehensive plan must be developed and a tool for directing land use must exist--comprehensive mapped or performance zoning. This is more or less true in each of the three population centers. El Paso may be the best prepared, with Ciudad Juarez a close second and Dona Ana County far behind.

5.1.1 El Paso

Although land use planning is accomplished at a consistently high level in El Paso, the implementation of the comprehensive plan is poor. There is an administrative separation between planning and zoning in El Paso and plans are only advisory. In El Paso County, zoning does not exist. (2) There has been a series of low income subdivisions developed in the county. Often they are without water or sewer service. (17)

5.1.2 Ciudad Juarez

In Juarez there is a strong tradition of local planning. The city has carried out local plans for a number of years, including comprehensive plans in 1984 and 1989. (3.4) Land use for large developments is controlled through a form of cooperative zoning between the developer and local government. Local government attempts to find the most profitable location for any development which will generate jobs and tax revenue, while minimizing the cost of infrastructure construction.

The major problem in Juarez is uncontrolled residential growth on the urban fringe-the colonias. This has occurred with increasing frequency and over large geographic areas. Colonias are beginning to appear south of the Santa Teresa-San Geronimo border crossing (southeast of the "A"-Figure 1). If the border crossing begins to draw significant amounts of traffic and the roadway is improved, without adequate land use control, it is expected that Juarez's colonias will spread in this direction at an increased rate. This would compound the infrastructure problems, because there are no municipal water or sewer services in this area.

5.1.3 Dona Ana County

Of the three population centers, there is no question that Dona Ana County is most at risk to experience serious land use problems from a successful NAFTA agreement. Currently, there are no plans or land use controls that could effectively direct industrial or commercial development and the resultant population increase that will occur in the southern portion of the county.

The last serious county land use plan was completed in 1967. (6) Zoning has been accomplished by special use permit on a case-by-case basis. The result has been a leap-frog pattern of uncoordinated land uses throughout the county.

In early 1993, a district judge declared the county's zoning ordinance unconstitutional in recognition of it vague character and arbitrary decision making process. He also strongly recommended that a new comprehensive plan be completed. (9) As of the summer of 1993, the county began to rewrite the zoning ordinance and was organizing to undertake a new comprehensive plan.

5. CONCLUSION

The North American Free Trade Agreement promises a great deal to the signatories of the treaty. Clearly, it will also impose costs. Not only will there be regional job losses and job gains, but as indicated in this research, there will be significant infrastructure costs. Although it is difficult to assess specific impacts on the region, it is clear that local governments in the central Rio Grande Valley are anticipating that NAFTA will increase border traffic. In addition, the need for essential systems like water and wastewater are at critical levels even without NAFTA.

There is some question whether this region can accommodate a large expansion in industrial production or population increases given the severe limitations on water and wastewater systems. In El Paso and Juarez, where planning has been effective, a strong link must be built into the systems that make the plans a reality.

In Dona Ana County, there must be a realization that it must become more aggressive in it land use planning and control efforts. The county is changing from one urban area with a rural hinterland to a multicentered urban landscape. Dona Ana County has a choice to either become responsible for the spatial organization of land uses or expect that the inefficient and leap-frog land use pattern will work against the positive effects that NAFTA may bring.

6. REFERENCES

1. Border Research Institute, 1993. "The Challenge of Planning Along the US-Mexico Border" (unpublished report). Las Cruces, NM: New Mexico State University, March 9.

2. Czerniak, Robert J., 1992. "Transportation and Land Use Planning in El Paso Texas: A Regional Analysis", Papers and Proceedings of Applied Geography Conference. V. 15, p. 39-47.

3. Estado de Chihuahua y h. Ayuntamiento de Juarez, 1984. Plan Director de Desarrollo Urbano, Ciudad Juarez, Chihuahua, November.

4. Estado de Chihuahua y h. Ayuntamiento de Juarez, 1989. Plan Director de Desarrollo Urbano, Ciudad Juarez, Chihuahua, November.

5. Federal Highway Administration, 1991. Intermodal Surface Transportation Efficiency Act. Public Law 102-240, Section 6015, December 18.

6. Harland Bartholomew and Associates, 1968. The Comprehensive Plan for the Urbanizing Portions of the Dona Ana County, New Mexico. Las Cruces, New Mexico. March.

7. Harris, Linda et al., 1990. Whose Water Is It Anyway? Las Cruces, NM: Arroyo Press.

8. Hufbauer, Gary Clyde and Jeffrey J. Schott, 1993. NAFTA An Assessment. Washington, D.C.: Institute for International Economics, February.

9. Jobe Concrete Products, 1993. Inc. v. Board of County Commissioners of the County of Dona Ana, County of Dona Ana, State of New Mexico, (Third Judicial District Court), May 18.

10. Loy, William, 1983. "Juarez Water Problems". Presentation at New Mexico State University, September.

11. Luckingham, Bradford, 1982. The Urban Southwest. El Paso, TX: Texas Western Press.

12. New Mexico Environment Department, 1992. Water and Wastewater Disposal Needs of Colonias in New Mexico. Santa Fe, NM, May.

13. Nickey, Laurence, 1992. "Third World Threats in a Modern Society". El Paso City-County Health District. El Paso, Texas. May 22.

14. Public Service Board of El Paso, 1992. Water and Wastewater Needs of Colonias in Texas. El Paso, TX.

15. Schuettler, Darren, 1992. "Farmers Fear Trade Deal: Draft Show No Safeguards for Quotas." The Financial Post March 28.

16. Texas Department of Highways and Public Transportation (TDHPT), 1991. "Projects Potentially Impacted by Free Trade Agreement Along Texas/Mexico Border". Austin, Texas. June 20.

17. U.S. Government Accounting Office, 1990. Rural Development: Problems and Progress of Colonia Subdivisions Near Mexico Border. Washington, D.C. November.

GRAPHIC NARRATIVES FOR EMERGENCY MAPPING

Mark Monmonier
Department of Geography
Syracuse University
Syracuse, NY 13244-1160

1. INTRODUCTION

Emergencies accompanying natural or technological disasters are among the most stressful and error-prone types of map use. The size and complexity of the problem, the dire consequences of mistakes, and pressures to act promptly obviate the more thoughtful, cautious and vigilant use of cartographic materials possible in facilities management and transportation planning. Designing maps for disaster response can be especially difficult because map users can vary widely not only in their experience in managing and carrying out evacuations and relief efforts but also in their knowledge of the area and the data as well as their skills in map reading and interpretation. (1) Without specific efforts to tailor presentations to both the stressful context of a crisis and the variable abilities of users, increased use of geographic information systems and electronic graphics is unlikely to meet the cartographic needs of emergency management agencies.

Graphic narratives offer a promising strategy for improving cartographic communication in a crisis. A multimedia solution that recognizes an inherent need for more than one map, graph, diagram, or text screen, the graphic narrative promotes interpretation and understanding through an ordered, coherent sequence of displays. This paper discusses the graphic script and other concepts of narrative graphics, reports on the Atlas Touring Project and the development and evaluation of two prototype graphic scripts, and explores the relevance of the project's findings to emergency management.

2. PRINCIPLES OF GRAPHIC NARRATION

Graphic narratives are an extension of the concept of a script developed by artificial intelligence experts Robert Ableson and Roger Schank, who observed standard sequences of questions, answers, comments, actions, and reactions in such task-oriented conversations as ordering a meal in a restaurant. (7) Schank defines the script as "a kind of play that we can engage in where our lines are prepared by a kind of general societal agreement, where we anticipate the lines of our partner in their likely place in the play and react accordingly." (6) It is necessary, of course, that scripts vary with circumstances as well as with the performers' backgrounds. Programming a computer to converse intelligently with humans thus depends upon a comprehensive library of scripts and a mechanism for identifying and adapting the right one. Although extending the notion of scripts to cartographic communication requires replacing much of the natural-language code of human speech with graphic symbols, a graphic narrative also depends for its effectiveness on the system's ability to anticipate the user's information needs by presenting relevant displays in an appropriately intelligible sequence.

2.1 GRAPHIC SCRIPTS AND GRAPHIC PHRASES

The Atlas Touring Project is attempting to integrate Schank's concept of a script with the notion of a twenty-first century electronic atlas able to offer customized tours of a geographic database. A guided tour can be customized in a variety of ways, for example, to provide an overview of the data, to address concerns about data quality, to explore a particular explanatory relationship, or to identify and describe geographic patterns the viewer might find interesting. In principle, a tour can respond not only to the user's inherent need for multiple graphic views but also to preferences for specific measurements and designs as well as concerns about particular regions and patterns.

A key concept is the graphic script, defined as a purposefully ordered sequence of maps, text screens, diagrams, statistical graphs, and other images selected and integrated to address a specific problem or query. (4) A shorter, more focused and fundamental sequence is called a graphic phrase. As an example, a graphic phrase useful in the early stages of a flooding emergency might generate a series of maps identifying roads likely to remain passable as the river rises one foot, two feet, and so on above flood stage. Roads that remain passable longer provide relatively reliable evacuation routes, and areas severed early from the overall road network would require early evacuation. This graphic phrase might be an appropriate component in a more comprehensive graphic script useful in identifying and describing neighborhoods and isolated residents at risk and in developing an evacuation plan.

Work to date has focused on the design of arguably logical sequences of automatically generated dynamic maps and statistical graphs. Two prototype scripts developed in 1991 proved useful in identifying a number of principles for composing graphic scripts. (5) The first prototype, called the "Correlation Script," examines the relationship between a dependent and an independent variable in a three-act play that progresses from a detailed description of each variable individually in the first act, to an exploration of their covariation in the second act, to the introduction of a regression model and an examination of residuals in the third act. Fill patterns for bar graphs, choropleth maps, and other displays reinforce the identity of each variable with a distinctive, highly contrasting signature hue--red for the dependent variable, blue for the independent variable, and magenta for the residuals from linear regression. Each act is subdivided into scenes, and each scene presents a variety of closely related text, maps, and graphs. To minimize the distraction of needlessly complex transitions, the script adds one new element at a time, and its various text windows, maps, and statistical diagrams have standard screen locations. As demonstrated by research on reader expectations, linking new material to old information promotes coherence and clarity. (2) To encourage understanding, a text screen defines each variable before its introductory scene in the first act, and a small text window introduces each dynamic sequence by announcing its point, or purpose.

At various places in the script, graphic phrases orchestrate an analysis of geographic patterns by guiding the viewer through a systematic examination of a linked map and graph. Called "canvass" sequences, these phrases follow a <u>narrative thread</u> based on places grouped by data category or region and highlighted in turn with blinking symbols or a temporarily different color. In the Correlation Script a typical canvass-by-category sequence progresses from the highest to the lowest data category, whereas a canvass-by-region sequence moves roughly from east to west. The second prototype script, the "Historical Script," includes canvass sequences with a third kind of narrative thread: time. Designed to explore spatial-temporal data using complementary absolute and relative measurements, this second Atlas Touring prototype demonstrates the potential of dynamic graphics for exploring the stability of rates of change by varying the length, starting point, and direction of movement of time periods. All three narrative threads--magnitude, region, and time--are relevant to emergency management. Natural processes, evacuation plans, and remediation processes have their own narrative threads, mostly based on time.

2.2 FOCUS-GROUP EVALUATION

Focus-group interviews conducted in 1992 provided information useful in refining principles for composing graphic scripts as well as in making the prototypes more effective. An experienced focus-group <u>facilitator</u> moderated the sessions--however thick-skinned, no designer should expect fully frank comments from viewers coaxed or dared to criticize his work. The facilitator collaborated with the principal investigator in developing a focus-group <u>protocol</u> that integrated discussion issues and "probe" questions with two software demonstrations. The first part of the group interview consisted of an approximately 11-minute demonstration of the Correlation Script, narrated by the facilitator, followed by a moderated discussion of the script's informativeness and coherence as well as its good points and bad points. In the second portion, designed to solicit more general opinions about customization and user interaction, the facilitator demonstrated two dynamic spatial-temporal maps from the Historical Script and an interactive version of selected graphic phrases from the Correlation Script's first act. To assure thoroughness and uniformity, the facilitator and the principal investigator rehearsed both the protocol and the facilitator's oral presentation of the demonstrations. Each session lasted about one hour.

To make the results meaningful, the principal investigator recruited 26 participants with experience in map design, cartographic analysis, or human-computer interaction. The facilitator conducted group interviews at four sites: Syracuse University's School of Information Studies (IST), the U.S. Geological Survey's National Mapping Division (in Reston, Virginia), the National Geographic Society's Cartographic Division (in Washington, D.C.), and the GIS applications unit at IBM Corporation's research and development center in Kingston, New York. The four groups ranged in size from five to ten. The IST group consisted of faculty and doctoral students; participants in the other three groups were full-time employees, most with more than five years of professional experience. For each session, the facilitator and the principal investigator studied the audio recordings independently, took notes, and collaborated in preparing two lists: one that identified consensus opinions and illustrated participants' views with revealing quotes, and a second that collected suggestions potentially useful in the next design cycle. Unlike the primary list, this second list included the principal investigator's and the facilitator's own ideas stimulated by the recorded discussion.

Like other qualitative data-gathering methods, the group interview can be highly subjective as well as nonrigorous. Its usefulness in developing products or concepts reflects the synergetic setting of a small group of appropriately selected participants encouraged to think about specific issues and to react to and amplify each other's insights and opinions. (3) Although investigators can minimize bias in the wording of discussion questions and by employing a comparatively neutral facilitator, interpretation of the results is innately imprecise. Systematic content analysis used in marketing research to promote consistency is unwarranted with a small number of participants surveyed in a few groups by a single facilitator. Confirmatory statistics are irrelevant as well because group interviews are intended to reveal a broad range of attitudes and opinions, not to estimate a population's characteristics from measurements collected for a small, desirably random sample. Nonetheless, consistency in the opinions registered across the four groups provided a rational basis for confidence in conclusions drawn from the taped discussions.

The focus groups were revealing. Most participants found the graphic scripts engaging, correctly detected salient patterns in the data, and appreciated the underlying graphic logic. They understood the signature hues but objected to the visually harsh blinking symbols used to highlight portions of some maps and statistical graphs. Participants also recognized the need for a voice-over or oral narrative similar to the facilitator's running commentary--although screens with text might appropriately precede a dynamic sequence of graphics, viewers cannot easily watch a moving display and simultaneously read titles, labels, and other accompanying text. Synthesized or recorded speech will become an important component of visualization support systems.

Of greater significance, participants unanimously wanted to control the pace of the presentation--once started, the prototype scripts run without interruption to their final scene, but viewers desperately wanted to slow down, stop, back up, and repeat selected portions. Although most agreed that a noninteractive, "closed" graphic script could be useful as a presentation medium or as a training tool to introduce the individual user to a new dataset, region, or analytical technique, participants felt strongly that efficient learning and understanding depend upon direct interaction with the data. In short, once introduced to a tool or data set by a training script, the experienced user demands full control. Participants also recognized that an intelligent visualization support system able to anticipate an analyst's or a manager's needs by providing or suggesting the next most appropriate display could be valuable to both neophytes and experienced users. Even though the context of the Atlas Touring prototypes and the group interviews was hypothesis generation and data analysis, these findings are relevant to emergency management.

3. OPENNESS, INTERACTION, AND NARRATIVE MENUS

Graphic narratives need not be as closed as the prototype Correlation Script. A rudimentary yet effective way to make Atlas Touring scripts more "open" would be to add pause points after all graphic phrases, at appropriate intermediate places within some graphic phrases, and at all text screens. At these points in a graphic narrative, the system would pause and await the viewer's command either to repeat the preceding view or sequence, or to continue immediately to the next. A second, more open modification is to provide interaction points, at which the viewer may manipulate the current graphic phrase. Of course, the system might also provide a "player-piano mode" to let the user run a graphic script as a closed presentation.

A more flexibly open strategy would provide access to the script's text blocks and graphics phrases through one or more narratively organized menus. At the lowest level of menu control, all graphic phrases and text screens might be linked to a set of pull-down menus. For the Correlation Script and other narratives organized like a play, a menu bar at the top of the screen might present the three acts as a group of adjacent pull-down menus, arranged in sequence from left to right. Scenes or graphic phrases ordered from top to bottom within each menu would help the viewer recreate the script in its intended order as well as jump back freely to a particular segment.

In a different menu configuration, a command window could list in narrative order a series of short descriptors representing the script's text blocks and graphics phrases. In its fully automated, player piano mode, the system would progress down this list, highlighting individual descriptors in the command window while presenting the corresponding text block or graphic sequence in the view window. A two-dimensional version is the toolbox, a grid of cells containing icons or acronyms organized sequentially by row or column. If the narrative has branch points at which the viewer must choose among two or more optional sequences, the system might present the menu as a flow diagram useful as well for highlighting progress. To offset the awkwardness of making frequent manual choices from pull-down menus, command windows, graphic toolboxes or flow menus, a "next" key such as the keyboard's downward arrow might allow the user to conveniently advance to the narrative's next logical step. In this case, the menu only need appear when the user reaches a branch point or decides to consider another path.

Whatever its degree of openness, a graphic narrative is a statement of a preferred or canonical sequence of views or techniques. The statement might be a simple sequence or a complex web, and its basis might be logic, common practice, administrative mandate, or a knowledgeable author's pragmatic instincts. Because narrative graphics is a new concept, cartographers and information experts have little if any experience in designing or detecting canonical graphic sequences--indeed, the classic cartographic problem is the single optimum map, not the pragmatically effective set of complementary maps and graphs. Yet if Schank and Ableson are right, attempts to identify and formalize graphic scripts might reveal surprisingly little variety among demonstrably successful experts, credentialled or otherwise. In emergency management, of course, these canonical sequences can vary considerably by type of hazard or disaster. For example, maps and other graphics useful in dealing with a severe winter storm should differ significantly, if not dramatically, from emergency graphics for coping with a nuclear accident--and determining exactly how different might be useful in identifying an amalgamated script appropriate, however unlikely, for a nuclear emergency during a heavy snowstorm.

4. USER PROFILES, DISASTER PROFILES, AND AREA PROFILES

The success of graphic narratives in emergency management will depend upon <u>user profiles</u>, that is, files describing each user's interests, experience, training, and preferences for specific types of presentations, details, and symbols. This information would allow a visualization support system to customize graphic narratives for individual users--delivering background information important to those unfamiliar with the area, the data, and the problem but not needlessly antagonizing those users more fully aware of both the situation and an appropriate response. Especially relevant is the user's experience in disaster management and cartographic competence. Developed during a query session when the viewer first accesses the system, the user profile could be updated at the end of the session, saved for later use, and updated as needed thereafter. User profiles could also be evaluated and modified during periodic training sessions. In addition, residential information in the user profile could help the system promote morale during an emergency by assuring workers of the safety of their own families--and advise officials about workers likely to perceive added stress as a result of family members still at risk.

Other metadata useful in customizing a graphic narrative include the disaster profile, the data profile, and the area profile. A catalog of <u>disaster profiles</u> could not only guide a visualization support system in adapting a master script to a wide variety of emergencies but also help program managers evaluate the adequacy of their databases, their equipment, and their personnel resources. Similarly, a <u>data profile</u> would help the system brief the user on the assets and limitations of the data, and thereby discourage unwarranted inferences. And an <u>area profile</u> would inform the user about significant landmarks and reference features, local slang, nonofficial place names, and other unique occurrences with which state and federal officials might not be aware. In addition to supporting training sessions and actual emergencies, these profiles are valuable planning tools, useful for informing and guiding periodic evaluations of geographic information, personnel, and equipment.

5. CONCLUDING REMARKS

However unattainable and unrealistically futuristic graphic scripts might seem at present, graphic narratives are fundamental elements in the inevitable (if not imminent) "new cartography" of dynamic maps and visualization support systems. Implementation of graphic narratives, user profiles, and related concepts will be especially valuable in emergency management to assist users ignorant of the impacted area, the data, or the hazard, or with limited skills in map reading and interpretation. In addition, by promoting a more vigilant and conscientious use of maps, graphic narratives can offset the effects of stress during a crisis. Moreover, graphic narratives should prove useful in evaluating information resources, in developing evacuation and response plans, and in training local officials to deal with disasters.

There is little formal knowledge of canonical sequences of maps and map-reading tasks upon which graphic narratives should be based. Yet even though information systems are unlikely to implement graphic scripts for another decade, it is not too early for emergency management agencies to address issues of cartographic complementarity and informative sequencing. Assessment of information needs, particularly the front-line user's need for geographic information, can identify gaps in existing data that diminish coherence and understanding. Because they represent serious defects in traditional, map-based information systems as well as impediments to the adoption of graphic narratives, such gaps warrant immediate attention.

6. REFERENCES

1. Dymon, Ute J., and Nancy L. Winter, 1991. "Emergency Mapping in Grassroots America: a Derailment Evacuation Case Study." Geoforum. 22:377-389.

2. Gopen, George D., and Judith A. Swan, 1990. "The Science of Scientific Writing." American Scientist. 78:550-558.

3. Krueger, Richard A., 1987. Focus Groups: A Practical Guide for Applied Research. Beverly Hills, Cal.: Sage.

4. Monmonier, Mark, 1990. Atlas Touring: Concepts and Development Strategies for a Geographic Visualization Support System. CASE Center Technical Report No. 9011. Syracuse, N.Y.: New York State Center for Advanced Technology in Computer Applications and Software Engineering, Syracuse University.

5. Monmonier, Mark, 1992. "Authoring Graphic Scripts: Experiences and Principles." Cartography and Geographic Information Systems. 19:247-260.

6. Schank, Roger C., 1991. The Connoisseur's Guide to the Mind. New York: Summit Books, p. 83.

7. Schank, Roger C., and Robert Ableson, 1977. Scripts, Plans, Goals, and Understanding. Hillsdale, N.J.: Erlbaum.

USING GIS TO MEASURE RELATIVE RETAIL ACCESSIBILITY DIFFERENTIAL

W. Paul Gallimore
Environmental Systems Research Institute
Redlands, California 92373

1. INTRODUCTION

Locating successful and profitable retail store sites can be described as finding an area within a market in which demand for goods and services exceeds the supply. For any given time, there are areas within markets in which demand is greater than supply (i.e., under-served), supply is greater than demand (i.e., over-served), and areas where supply and demand are relatively balanced. These areas, or distributions of supply and demand, change in size and migrate through time due to changes in business cycles, population distributions, and demographics.

Smith (12), in an article observing the failure of a new community shopping center, refers to the "catch-all" for a project gone sour as "the numbers didn't work". He noted several location assumptions which did not materialize and eventually lead to the project's inability to be profitable enough to pay the rent. Foremost, population growth estimates for this site never materialized. Coupled with this lack of demand, competition (i.e., retail supply) increased nearby and the economy sagged. Therefore, conditions which existed when the project was planned changed by the time the project was built. By conducting a time-series analysis with varying supply and demand assumptions for a given market, perhaps better retail sites can be selected. Retail location decisions are difficult at best ; however, there are tools available to assist the analyst in conducting these studies.

Several methods and techniques have been developed by researchers in geography, marketing, and economics which attempt to quantify the retail location question. These spatial interaction models (SIMs) trace their history back to the 1930s with Reilly's work developing "the law of retail gravitation" (10) and are grouped into a category known as gravity models. In general, these models relate interactions between people and stores based upon distance and attractive mass (i.e., store size). Derived from and related to gravity models are spatial accessibility models, which include retail potential models.

Retail potential models in general attempt to measure access from sub-areas within a given market to retail facilities within that market. In theory, stores with greater access to their patrons will realize higher potential sales due to a higher probability of being visited. As with gravity models, accessibility models are based upon the concepts of customer distance and store attractiveness. As customers attempt to minimize travel costs through rational store choice, accessibility to such stores increases. Accessibility models, or some form of them, have been used in several previous service and retail location studies (3) (4) (11). Though these models are conceptually suited for assisting the analyst in making store location suggestions, they require extensive computations. In terms of data integration, technique integration, and concept integration, the analyst requires a tool which can put all the elements of a retail location model together.

Several recent advances in data processing can be applied to assist the locational analyst in conducting these difficult studies. One of these advances, which is applicable to retailing, is known as Geographic Information Systems, or GIS. Within a GIS, data not only have a magnitude and temporal component, they also have a spatial component. Through visual interfaces for input and output, GISs have the ability to display complex spatial data sets in a graphically clear manner. This functionality, in turn, provides new and different perspectives when analyzing data. As Moloney (8) noted, dynamic simulations can be performed to determine the impact of various strategies. Different locational responses such as store expansions or closures can be tested and evaluated. GISs make the task of data integration, data analysis, model building, and scenario generation more manageable. With this functionality comes the benefit of being able to generate more alternatives in less time.

In general, this research integrates retail supply and demand data from several time periods into a GIS (Arc/Info) which has the ability to perform spatial interaction modeling. Through this integration, a simplified retail accessibility potential model is developed which, in turn, is integrated into a prototype retail spatial decision support system (RSDSS). This prototype RSDSS is tested in a case study using basic simplified retail supply and demand data. Such a prototype could then be refined by the practitioner (analyst) and applied to a particular locational need.

Discount department stores located in Charlotte, North Carolina were chosen as the subjects of this case study. Specifically, this study is limited to analyzing the retail accessibility (relative disparity between retail supply and retail demand) of the following discount department store chains: K-Mart, Target, market newcomer Walmart, and the defunct chains of Zayre and Woolco. The time periods included in this study were 1980 and 1990. Scenario testing included changing retail supply (i.e., introduction of new competition), changing retail demand (i.e., population distribution), and changing the road network (i.e., road additions).

2. RESEARCH DESIGN

2.1 MODELING OVERVIEW

In defining the model for the prototype retail spatial decision support system, relative measures of supply and demand are developed for the study area by using the "entropy" form of the accessibility model. The study area is defined as a county and is divided into zones based on census block groups. Demand data for these block groups are collected or estimated for 1980 and 1990. Retail supply data are also collected for these times. Zonal distance separations, or travel cost, are calculated as travel time through a "sketch" network of roads within the county. Supply and demand data are added to this network through the use of "loading" nodes on the network. Supply and demand indices are normalized, and are used to generate surfaces which are compared to determine areas where supply is greater than demand, less than demand, and equal to demand. These surfaces are generated for 1980 and 1990 and compared.

2.2 MODELING ASSUMPTIONS

Several modeling assumptions are made when defining accessibility formulas, including the following: (1) accessibility from consumers to opportunities is a function of travel cost measured in travel time, (2) all individual consumers in a zone possess similar attributes and live at one point in the zone, (3) within the transportation network, there is one measure of travel cost between any pair of nodes, and, (4) actual shopping behavior is not being measured or predicted. The analyst should note that accessibility measures generated in this study do not reflect any information on opportunities outside the study area.

2.3 SPECIFYING THE DATA

Model specification includes: selecting a retail supply measure for the service or good being researched, selecting a retail demand measure for that service or good, and defining a transportation network upon which consumer trips are modeled. Retail supply is defined as department store square footage which are located within the study area. Retail demand is measured as census block group populations for a given year. Transportation network data required by the model include a set of links and nodes covering the study area. Travel speed data are used (i.e., speed limits in miles per hour) to develop network travel times.

Not all of the roads in the county need to be used for the modeling environment. Instead a "sketch" network can developed using only the roads of major significance within the study area. Typically these roads are interstates, highways, boulevards, arterials, and multi-lane facilities which are considered to be high-level network connectors. In building a sketch network all these high-level roads are kept in the TIGER road network database while low-lever roads are removed. The only inclusion of local low-level roads in the network occurred when one of these roads passed close to the centroid of a census block group (i.e., demand location) or when one passed close to the location of a past, present, or proposed discount department store (i.e., supply location).

To load the supply and demand inputs into the accessibility model, a set of "loading" nodes were located on the network. In general, a node on a network is either the end of a road segment or the intersection of two or more road segments. Nodes found midway along a road, or an arc, are known as "pseudo" nodes. For each store and each block group, a unique loading node must exist on the network in order to compute accessibility. Several techniques are available for attaching retail supply and demand data to their respective loading nodes.

2.4 SPECIFYING THE ACCESSIBLITY MODEL

The "entropy" accessibility formula was chosen as the model "engine" of the RSDSS. The basic entropy accessibility measure (equation 1) is specified to be inversely related to an exponential function of distance where lambda (λ) is the model's distance decay parameter or constant. Accessibility is defined as:

$$A_i = \sum_{j=1}^{n} \frac{W_j}{e^{\lambda (TT_{ij})}} \qquad (1)$$

where:

- A_i is the accessibility or potential at location i
- n is the number of locations in the study area
- W_j is the weight measuring the attractiveness of location j
- λ is the parameter of distance decay
- TT_{ij} is the travel time between locations i and j

In this form of the accessibility model, the total modeled accessibility or potential A_i at location i cannot exceed the total available opportunity measures W_j summed for all locations j. The weighting parameter of this model (W_j) can use retail attractiveness (e.g., store square footage) or a measure of retail demand (e.g., population). Each 1990 population loading node has a separate supply accessibility score and a separate demand accessibility score calculated by these formulas. Conceptually, by developing both a supply and demand component of retail location and comparing the two, a more complete model is developed.

2.5 SELECTING THE DECAY PARAMETER

Choice of appropriate values for the decay parameter to be used in calculating the accessibility measure is not easy. Selection of this parameter is known as calibrating the model and is a very critical part of the modeling process. Normal practice for calibrating travel studies includes the selection of a parameter value which provides the closest statistical fit between observed travel behavior and that predicted by the model (5). When empirical data is absent statistical techniques are used to calibrate the model. These techniques include examining the frequency distributions of the model's supply and demand accessibility scores, plotting the standardized supply and demand scores on a scattergram and observing the resulting patterns, and dividing the highest supply and demand accessibility score by the lowest and examining the resulting ratios.

Observed results from previous studies can act as a guide in parameter selection. Decay coefficients vary depending on retail activity (i.e., fast food, convenience stores, groceries, discount department stores, etc.) because trade areas vary by activity. Huff (7) noted different distance decay exponents for different types of shopping trips. He observed that larger estimated decay values yielded smaller time expenditures or shorter average trip lengths. Thus, the greater the decay, the more restrictive will be the scope of the trading area. A high decay coefficient would be used for an activity such as a convenience store whose primary trade area may only be extending out a mile or so. A slightly lower decay would be used for grocery stores whose primary trade areas extend out to perhaps 2 miles. An even lower value would be used for discount department stores and speciality stores where primary trade areas may extend to 4 miles and beyond. Conventional department stores and regional malls would have yet an even lower value, yielding much longer average trip lengths.

2.6 INTERPRETING THE RESULTS

After calculating both retail supply accessibility and retail demand accessibility for each of the block group loading nodes, the analyst needs to combine the supply and demand accessibility measures into a single expression yielding a spatially continuous index of relative accessibility. This index is a measure of retail market opportunity.

This research develops its index by using the statistical concept of the standard score. A standard score is simply the degree to which an observation's value differs from the mean of its distribution, measured in units of standard deviation (6). These measures are dimensionless, or independent of units used. By converting the demand and supply scores into standard scores, measured in common units, the two scores can then be compared. With both demand and supply in the same scale, the relative differences between the two, at a given location, can now be calculated. This difference, termed relative retail accessibility differential ($RRAD_i$), is determined by subtracting the z-score of retail supply accessibility (ZDA_i) from the z-score of retail demand accessibility (ZSA_i) as shown in equation 2.

$$RRAD_i = ZDA_i - ZSA_i \tag{2}$$

This retail accessibility differential index is a relative measure since there is no basis for determining what level of accessibility differential would be adequate for the study area. Such research lies outside the scope of this study and would make an excellent topic for future studies. Comparative analysis between retail supply and demand accessibility is facilitated by using this measure, and can easily identify areas where there is a significant disparity between retail supply and demand. Where the RRAD index is positive, demand for retailing is greater than its supply. A high relative retail accessibility differential for a given location indicates the latent need for retail function. Conversely a negative RRAD index indicates that the supply of retailing is higher than its demand. Locations where relative retail accessibility differential is close to zero indicates that retail supply and demand are in balance.

When demand is greater than supply, the analyst should look at population and income trends for that area before immediately concluding that this is the area to add an additional store. Several model scenarios should be run by varying supply and demand assumptions. What would be the effect of the competition adding a store close to the site under analysis. The analyst could vary the proposed store square footage to try and balance supply and demand for the proposed site. The model could be run for the site for different time periods to observe the relative accessibility trends and test a site for performance stability. GIS allows these multiple scenarios to be computed with relative ease once the initial model has been built and calibrated.

When supply is greater than demand, the same process of looking at supporting data should be used before making decisions based upon the model results (e.g., store closure). In addition to demographics, one can look at an existing store's level of sales, its management, it's interior and exterior image, and the advertising campaign for the area in which it is located. The competition's strengths and weaknesses can be observed. After thoroughly reviewing all available data, then a decision on what to do for a given store can be more rationally made.

2.7 VISUALIZING THE RESULTS

Since accessibility indices and differential scores are a continuous index, they can be used as inputs into 3-dimensional surface generation models. These surfaces help the describe the patterns of retail supply, demand, and relative over or under storing to the analysts and decision makers. Interpreting and visualizing model output is simplified using this technique. The method chosen to generate model surfaces was TIN due to its ease of use and straightforward approach in generating surfaces. TIN bases its predicted surface on fitting a flat plane to any three non-collinear data points (1). The underlying assumption requires that the input data point locations (i.e., block group loading nodes) represent critical values of the surface. Using a linear interpolation model, the original data values are retained in the resulting surface as its maxima and minima. By interpolating values along the TIN lines, data points of equal value are connected to form contours or isolines of equal value.

Once the TIN surface has been generated, there are several options available to the analyst for viewing and analyzing this surface. A TIN surface can be converted into a lattice or grid surface which is composed of several square cells (e.g., pixels) formed from a series of horizontal and vertical lines. These cells contain a z-axis data value interpolated from the TIN surface. Since each cell in the grid represents a data value of a continuous data surface, 2-dimensional planar maps of this data can be viewed using different colors or shades to represent different data values. For example, a red-to-blue color ramp could be used to view accessibility indices with high accessibility values represented by bright reds and low accessibility values represented by blues. Purple colors would be interpreted as areas with medium accessibility. These "grid-painted" maps could then be draped on top of TIN surfaces to produce 2 1/2-dimensional or "pseudo" 3-dimensional displays.

2.8 SELECTING THE GIS SOFTWARE

ARC/INFO revision 6.1 was selected as the GIS software to be used in building the prototype retail spatial decision support system. It was selected because of the following functionalities: spatial interaction network modeling, surface generation and display, and customizable interface development. Both the gravity and accessibility spatial interaction models were contained within the software. By having all the functionality contained in one package, the development of a prototype application was easier to construct.

3. CASE STUDY - CHARLOTTE DISCOUNT DEPARTMENT STORES

3.1 DATA

Mecklenburg County, North Carolina was the geographic area selected for the case study. Charlotte, the county seat, is a mid-sized emerging urban center with a population of approximately 400,000. Serving as the financial, manufacturing, and service center of the region, Charlotte is surrounded by several bedroom communities which supply commuting workers and consumers. In 1980, there were 15 discount department stores within the study area. By 1990, this number had dropped to 12. To observe these market changes, this study modeled discount department store accessibility for 1980 and 1990.

Discount department stores were chosen to be used as the model's retail supply data and included the following chains: Kmart, Target, Wal-Mart, Woolco, and Zayre. Data sources for store locations included the yellow pages of the 1990 telephone directory (13) and the Charlotte City Directory (9). Store size in square feet was approximated by visiting store sites and measuring the length and width of the buildings using a rolling wheel measure.

Simple population counts at the block group level were chosen to be used as the model's measure of retail demand. These data were collected from the 1990 Census PL-171 summary data files and aggregated for the 399 census block groups. Of these 399 block groups only 398 were used due to the isolation of one block group from the county by Lake Norman. Population counts were obtained for 1980 and 1990 census tracts from the Charlotte Chamber of Commerce (2) and 1980 counts were estimated for 1990 tracts. From these data, 1980 population counts were estimated for 1990 block group boundaries. The 1990 Census TIGER files provided census geography boundaries (i.e., polygons) for census tracts and block groups.

The model's network data were developed using the 1990 U.S. Census TIGER files for Mecklenburg County. Major highways and arterials with key connector roads were pulled from the complete county road network using the TIGER functional class code (CFCC) and compiled into the simplified modeling sketch network which consisted of 3762 links and 2676 nodes. Population and "loading" nodes (i.e., nodes which passed close to block group centroids) were chosen using roads which existed in 1980 and and 1990. Store loading nodes were located near the store's street address. Network speed limit data were collected for the major roads (i.e., interstates, federal and state highways, and local arterials) within the study area by windshield survey. For those roads which were not driven, speed assumptions were made made on road classification.

3.2 CALIBRATION OF MODEL

After loading the retail spatial decision support system (RSDSS) model with supply and demand data, model calibration was performed by adjusting network impedance travel time and by adjusting the model's distance decay parameter. 1990 data was used to calibrate the model.

Calibration of the sketch network's travel time impedance measure was conducted by simulating drive time for several trips over the model's road network, and comparing the simulated drive times to the actual observed drive times for those trips. From these observations a formula was defined to calculate travel time for the network.

Once the model's network impedance was calibrated, the model's accessibility decay was calibrated by varying the value of this parameter, holding all other model inputs (i.e., supply, demand, network) constant to 1990 values, and observing the resulting model's outputs using simple descriptive statistics. Initial parameter values to be evaluated ranged from 0.1 to 1.0 and were chosen based upon previous work found in the literature (5). A decay parameter of 0.28 was chosen, for the model for several reasons. This parameter produced a more normalized frequency distribution for demand (i.e., z-demand) accessibility than the other tested values. The difference between the absolute value of the maximum z-demand value minus the absolute value of the minimum z-demand value was best balanced around the mean using this value.

3.3 ANALYSIS AND INTERPRETATION OF RESULTS

After calibrating the model, several model tests and scenarios were generated by varying the model's supply, demand, and network inputs. Network comparison models examined the effects of using different road networks to calculate retail accessibility. Market models examined retail accessibility for specific time periods (i.e., 1980 and 1990). Model parameters and results are summarized in Table 1. Model outputs (i.e., retail supply, retail demand, and relative retail accessibility differential) were additionally assessed by using both 2-dimensional grid-painted surfaces and 3-dimensional grid-draped oblique surfaces. For both these maps, light areas represent areas with high values and dark areas conversely represent areas with low values.

The base model used in this study was the model that used the 1990 population, store square footage, and road network as inputs (M1990). This model was used to determine the decay parameter which was held constant for all the models. The resulting supply and demand accessibility surfaces demonstrates some similarities in their patterns. Areas of high accessibility were found in the center of the study area with accessibility decreasing towards the county periphery. These results were expected due to both data distribution (i.e., supply and demand values are lower along the county border), and model performance (i.e., boundary loading nodes have an inherently lower accessibility due to less access to the network).

When the supply and demand surfaces are combined arithmetically by subtracting supply from demand, maps of relative retail accessibility differential are produced. Areas where demand is greater than supply are located in both the eastern part of the county and in the central and south central areas of the city (Figures 1 and 2). Areas where supply is greater than demand (i.e., relative retail accessibility differential is low) are found near existing store clusters. The area where retail accessibility differential is lowest is located along the South Boulevard corridor in the southern part of the county.

Once the base model was computed, several tests were conducted to observe the effects of network improvements upon retail accessibility and the derived retail accessibility differential. These tests held the base model's 1990 supply, demand, and decay parameter inputs constant while substituting road network data. The N1980 model performed as expected by yielding lower aggregate supply and demand scores than the N1990 and the N2000 models. As network efficiency increases, one would expect aggregate accessibility to increase. This relationship can be used to test the network's data integrity. If, after network improvements, aggregate accessibility declines, then most likely, network coding errors are present. Another observation made from these tests show that the overall supply to demand ratio is increasing with network efficiency. This ratio increase indicates that aggregate supply accessibility is benefiting slightly more than aggregate accessibility as these network additions improve overall county accessibility (Table 1).

One of the goals of this study was to integrate temporal changes into the retail spatial decision support system. This temporal market change was observed by using 1980 retail supply, retail demand, and road network data in the model while holding the decay parameter constant. This model was designated M1980 and was compared to the 1990 market model (M1990). The resulting overall supply to demand ratio for 1980 (1.1088) is almost two times that for 1990 (0.6363). The net loss of retail supply (i.e., store square footage) and an increase in retail demand (i.e., population) between 1980 and 1990 resulted in the much lower ratio in 1990.

4. CONCLUSIONS

4.1 CASE STUDY OBSERVATIONS

This case study has revealed that a site for a future discount department store may exist in the eastern part of Mecklenburg County due to the increasing accessibility differential measured in this area. This conclusion appears to be shared by one of the discount department store chains (Wal-Mart) with the the recent construction of a new store in this area. Determining the proper size of the store and exact location of the site requires further research and testing. After identifying areas of possible store placement, the analyst would then extend the study to reviewing other pertinent site locational data such as demographic trends, zoning ordinances, land availability, and traffic projections before concluding to locate a new store at a given site.

4.2 MODELING ISSUES

This research provides an example of the development of a prototype GIS retail spatial decision support system. As a research and demonstration project only, this research was not intended to be a strategic decision making system. Such a system would have to be implemented by the practitioner who has had experience weighing all the required data for making specific location decisions. The model developed in this study only identifies relative gaps between retail supply and demand based upon network accessibility and acts as a "screening" process to assist the locational analyst in selecting candidate sites for further investigation. This model does not address threshold identification for market entry and market exit, nor does it make predictions on retail supply absorption rates or site performance.

Care needs to be taken when specifying the model (i.e., in making model assumptions). One should be somewhat familiar with the retail segment being modeled before selecting the model's accessibility formula and decay parameter. If time permits, several alternate tests should be conducted using several different accessibility formulas. These results should be compared to determine which formula best describes the retail category being modeled. Previously developed models are not transferrable from market to market; however, the procedures and techniques can be replicated in other markets. Elements such as travel time impedance and distance decay parameters have to be determined for each individual market in which the model is used. Different decay parameters are needed for different retail activities (i.e., department stores, discounters, speciality shops, groceries, convenience stores, etc.).

The analyst should exercise caution when analyzing retail accessibility differential as measured in this retail spatial decision support system. The display of the model's results can mislead the analyst due to several factors. The RSDSS will always locate areas where supply and demand diverge and display these differences using the same shading technique (i.e., the highest differential scores are shaded light gray while lowest scores are shaded dark gray). These highlighted areas may not be significant enough to justify changes in store alignment. A given market could be balanced and the differences between supply and demand may be insignificant. Overall, the model appears to be a good screening technique for identifying areas for further research and investigation by the location analyst.

4.3 GIS ISSUES

GIS serves several roles when one conducts retail market research. These roles include serving as both an integrator of data and of analytical techniques and as a scenario generator. Data from a variety of sources and scales is integrated into larger and denser study areas. A recent development in the GIS industry is the availability of purchasing pre-fabricated data sets in common GIS formats. Preprocessed data exists at a variety of scales and study area sizes for several locational model inputs. When using preprocessed data, the analyst needs to know the methodology of how the data were captured or created and any assumptions made during this process.

GIS is evolving in sophistication of tools, models, and statistics available for use in addressing more complex models. An analyst still needs a good foundation on applying basic location research techniques and interpreting their results. The analyst should be trained in how to properly apply GIS as a tool when conducting retail location analysis, not a GIS technician without location analysis background. Within this framework, there is plenty of room for GIS to be used in conducting retail location research.

5. REFERENCES

1. Branson, Wendy Saxton, 1989. "The Role of Point Data in GIS." The Operational Geographer 7 (4): 9 -12.

2. Charlotte Chamber of Commerce, 1991. "Charlotte Mecklenburg Census Tract Data Population 1980 - 1990." Information, Research from the Charlotte Chamber, Charlotte 1 - 4.

3. Cullen, Bradley T., and Laurence Spear, 1985. "Retail Coverage and Market Equilibrium: The Case of Food Stores in Albuqurque, New Mexico." Papers and Proceedings of Applied Geography Conferences 8: 258 - 272.

4. Guy, Cliff M., 1983. "The Assessment of Access to Local Shopping Opportunities: A Comparison of Accessibility Measures." Environment and Planning B: Planning and Design 10: 219 - 238.

5. Guy, Cliff M., 1986. "Accessibility Measures in Urban Research: An Application to Health-care Location Problems." Perspectives in Urban Geography: Models in Urban Geography Part-II (Mathematical) 4B: 647 - 674.

6. Haake, John H., 1981. "A Technique for Retail Strategy Planning." Paper presented at the Association of American Geographers annual meeting, April 21, 1981.

7. Huff, David, 1964. "Defining and Estimating a Trading Area." Journal of Marketing 28: 34 -38.

8. Moloney, Terance, 1989. "A Case Study using a Geographic Information System in Food Retailing." The Operational Geographer 7 (4): 23 - 27.

9. R. L. Polk and Co., 1980. Charlotte City Directory 1980. R. L. Polk and Company.

10. Reilly, William J., 1949. Methods for the Study of Retail Relationships, Austin, University of Texas Press.

11. Schultz, Ronald R., 1975. "A Space Potential Analysis of Physician Location." Proceedings of the Association of American Geographers 7: 257 - 259.

12. Smith, Doug, 1991. "Shopping Center Built on Dreams Collides with Slow-Growth Realities." The Charlotte Observer (July 31) 1D, 6D.

13. Southern Bell, 1990. Charlotte, N.C. Phone Directory. BellSouth Advertising & Publishing Corp.

TABLE 1
SUMMARY OF ACCESSIBILITY RESULTS

Model Name	Model Supply	Model Demand	Model Network	Decay Parameter	Total Supply Accessibility	Total Demand Accessibility	Supply / Demand
N1980	1990	1990	Net.1980	0.28	5276390	8294985	0.6361
N1990	1990	1990	Net.1990	0.28	5382034	8458779	0.6363
N2000	1990	1990	Net.2000	0.28	5443611	8482347	0.6418
M1980	1980	1980	Net.1980	0.28	8095220	7301104	1.1088
M1990	1990	1990	Net.1990	0.28	5382034	8458779	0.6363

FIGURE 1
1990 RELATIVE RETAIL ACCESSIBILITY DIFFERENTIAL

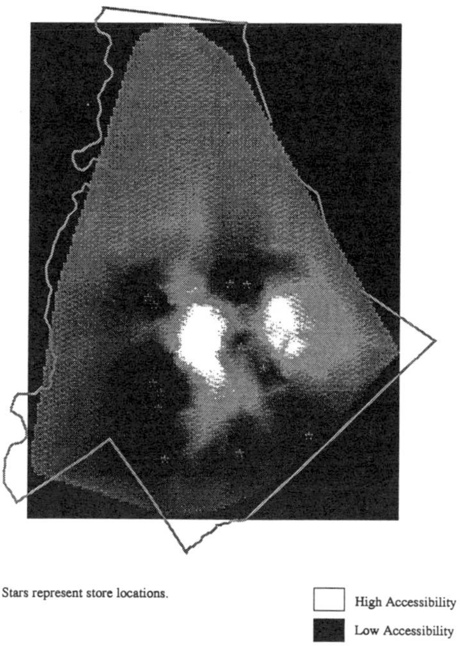

Note: Stars represent store locations.

FIGURE 2
1990 RELATIVE RETAIL ACCESSIBILITY DIFFERENTIAL, 3D VIEW

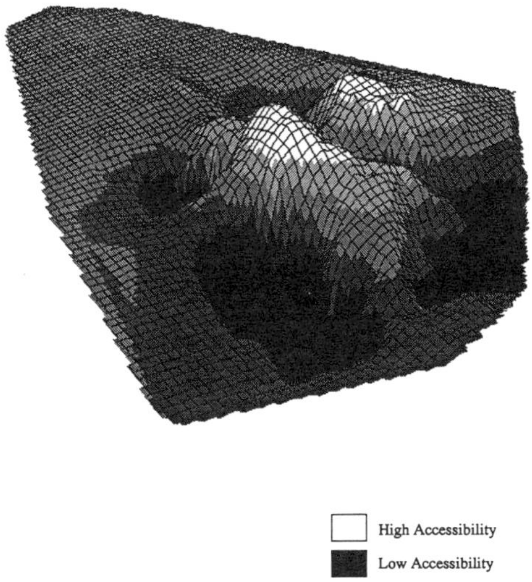

LOUISIANA LOTTERY PLAY BEHAVIOR AS A FUNCTION OF GEOGRAPHIC AND DEMOGRAPHIC VARIABILITY

Philip S. Kopel
Kopel Research Group, Inc.
Sharon, Massachusetts

1. INTRODUCTION

In recent years state governments have found that Lotteries are a good vehicle to increase revenue without the unpopular political repercussions of increasing tax rates. Lotteries have effectively become multi-billion dollar businesses, faced with similar decisions as are many other large corporations. These include developing effective target marketing approaches to maximize revenue while minimizing excess costs.

One research technique that has been utilized to help maximize revenue is a baseline telephone survey. The research provides the Lottery with the ability to index lottery game play state averages. The data is also cross-tabulated by demographic and frequency of Lottery games played. The Lottery can then utilize this information to optimize lottery agent locations, advertising campaigns and new lottery products.

2. SURVEY METHODOLOGY

This paper is based on a telephone survey that was conducted in Louisiana providing 1000 completed interviews (1). This sample size provided for an error factor of ± 3% at the 95% confidence level.

All respondents were at least 18 years of age, and the age structure of those interviewed was designed to correspond with the statewide age structure.

A 50/50 male to female ratio was maintained.

A stratified random sample was employed to create survey results which were representative of the population of the state. The sample was designed to match the state population at the Parish level of geography. A total of 8,000 telephone numbers were purchased in 50 replicates of 160 telephone numbers. As interviewers telephoned from the geographically stratified random sample, they maintained geographic representation continually to that of the state population distribution.

The stratified random controls were especially important to maintain in Louisiana since the survey is a snapshot in time in 6 month intervals (2) and comparisons over time need to be precise. Further, there are demographic differences as a function of geography and these need to be held in control.

3. DEFINITIONS

Scratch Tickets are games that have a variety of game themes and artwork. Players play these games for their "instant win" play style. They scratch off a play area on the ticket with a coin or similar object. Players win various amounts by matching a specified criteria such as matching 3 of a kind, matching a key number or adding up a score.

Lotto is a game that contains a multi-million dollar jackpot. This game is classified as on-line in that it is tied to a central computer and the numbers that a player picks are maintained in the computer. Since players pick their own numbers and other players could share their same number, the jackpot payout amount is pari-mutuel. The lottery draws the numbers live on television once per week. If nobody has all of the numbers, the jackpot "rolls over" to the following week and continues to build until there is a winner.

Pick 3 is a daily on-line game. In this game a player picks a number from 0 to 999. The Lottery picks a 3 digit number each night. If the player matches the Lottery's number he/she wins the top prize amount which is fixed at $500.

4. LOUISIANA LOTTERY GAME PLAY PROFILES

Approximately 70% of adult Louisiana residents have tried at least one of the lottery games. Approximately 60% of adult Louisiana residents are current players of at least one lottery game.

Each Lottery game varies in its play behavior and appeal due to the expected play frequency, jackpot amount, prize payout structure and overall odds of winning. One of the objectives of baseline research is to identify the differences in lottery play behavior by lottery game type as follows:

4.1. SCRATCH PLAYER BASE

Almost 60% of adult Louisiana residents have tried the Scratch game.

Among those market segments who have tried the Scratch games at a higher level than the percent of the state are those 18-34 year old (65%), with Household incomes $30,000 or more (63%), and Catholic (65%).

Among those market segments who have tried Scratch at a lower level than the state's percentage are those above 55 years of age (40%), Household incomes below $30,000 (55%), Asians (53%), and Protestants (53%).

Approximately 40% of adult Louisiana residents are current Scratch players (have played the Scratch tickets in the past month).

Almost 69% of those who have tried Scratch are current players.

The overall frequency of Scratch ticket play is as shown in figure 1.

Figure 1

FREQUENCY OF SCRATCH TICKET PURCHASE

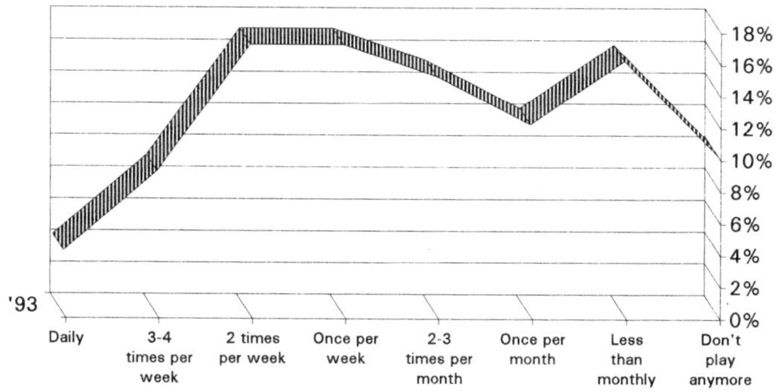

Almost 30% of all Scratch players play greater than once per week.

(1) Approximately 32% of the males who play Scratch, play greater than once per week compared to 28% of the females.

(2) About 46% of Blacks who play Scratch, play greater than once per week compared with only 24% of Whites.

(3) Only 14% of Protestants, who play Scratch, play greater than once per week. Similar to the state percentage, 36% of Baptists and 32% of Catholics play greater than once a week.

(4) Approximately 38% of the 18-34 age group play Scratch more than once a week.

(5) There is little difference among Income, or Education.

Those purchasing slightly more tickets than average each time they buy Scratch tickets are Females (3.0 tickets), 35-54 years of age (3.1 tickets), those with a High School or less education (3.0 tickets), Blacks (3.2 tickets) and Catholics (3.1 tickets).

Those purchasing slightly fewer tickets than average each time they buy Scratch tickets are males (2.9 tickets), college or greater education (2.86 tickets), Asians (2.5 tickets) and Protestants (2.6 tickets).

The remainder purchase tickets at about the average rate.

4.2. LOTTO PLAYER BASE

Almost 67% of adult Louisiana residents have tried the Lotto game. Among those market segments who have tried Lotto at a higher level than the percent of the state are those 35 -54 years old (73%), incomes $30,000 or more (73%), Asians (87%), and Catholics (78%).

Among those market segments who have tried Lotto at a lower level than the state's percentage are those 55 or older (53%), incomes below $30,000 (62%), Blacks (66%), and Baptists (59%).

Approximately 57% of adult Louisiana residents are current Lotto players. Almost 86% of those who have tried Lotto are current players. Approximately 68% of those who have ever played Lotto, play it at least once per week (every draw) as shown in figure 2.

Figure 2

FREQUENCY OF LOTTO PURCHASE

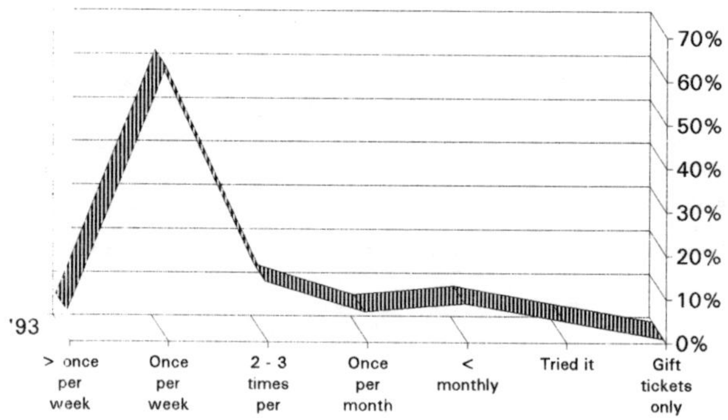

The average Lotto player buys 3.9 Lotto tickets each time he/she plays.

Those purchasing slightly more tickets than average each time they buy Lotto tickets are Males (4.1 tickets), 35-54 years of age (4.0 tickets), Household Incomes over $30,000 (4.2 tickets), Whites (4.0 tickets), Asians (4.4 tickets), Catholics (4.0 tickets), and Protestants (4.1 tickets).

Those purchasing slightly fewer tickets than average each time they buy Lotto tickets are Females (3.7 tickets), those 55 years of age or more (3.6 tickets), Household incomes less than $30,000 (3.8 tickets), Blacks (3.5 tickets), and Baptists (3.6 tickets).

The remainder purchase at about the average rate.

4.3. PICK 3 PLAYER BASE

To date 19% of the adult Louisiana residents have tried Pick 3. Among those market segments who have tried Pick 3 at a higher level than the percent of the state are, Males (20%), those with incomes $30,000 or more (21%), Blacks (26%), Asians (27%), Baptists (21%), and Catholics (20%). Among those market segments who have tried Pick 3 at a lower level than the state's percentage are those age 55 or more (11%), Whites (16%), and Protestants (15%). Only 11% of adult Louisiana residents are Current Pick 3 players. Current Pick 3 play frequency is presented in figure 3.

Figure 3

FREQUENCY OF PICK 3 PURCHASE

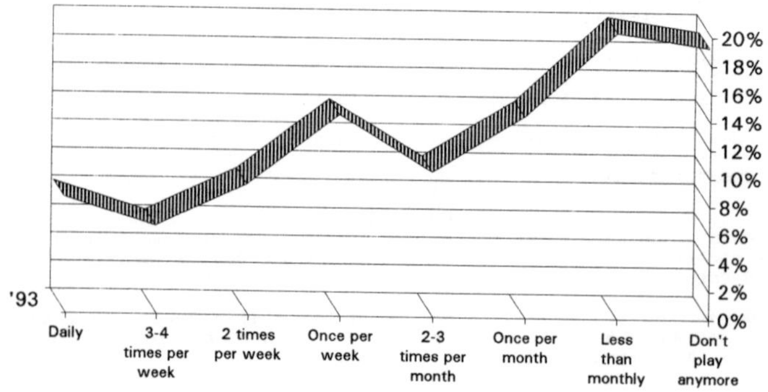

Less than 1/4 (23%) of all Pick 3 players play greater than once per week. Those segments that play Pick 3 more often than once per week at a slightly higher percentage than the state average are: those of ages 55 or greater (33%), household incomes below $30,000 (35%), those with education of High School or less (30%), Blacks (30%), and Protestants (31%). Those segments that play Pick 3 greater than once per week at a lower percentage than the state average are those ages 18-34 (20%), household incomes of $30,000 or more (9%), College educated (13%), Whites (19%), and Baptists (21%).

5. COMBINATION LOTTERY GAME PLAY BEHAVIOR

Many players of Lottery games play a combination of games. In fact, it is important that the Lottery merchandise the Lottery games to facilitate multiple game play. The following table presents the Lottery combination game play behavior by percent by demographic group as follows:

	Ever Played Any Game	Current Players Surveyed	Scratch Only	Scratch & Lotto Only	All Scratch Lotto & Pick 3 Games	Lotto Only	Lotto & Pick 3 Only	Scratch & Pick 3 Only	Pick 3 Only*
TOTAL	71%	100%	6%	43%	14%	33%	3%	1%	--%
Male	70%	100%	4%	45%	15%	32%	3%	1%	***
Female	72%	100%	8%	42%	13%	34%	3%	1%	***
White	70%	100%	5%	45%	11%	36%	3%	0%	***
Black	71%	100%	8%	39%	23%	26%	3%	2%	***
High School or Less	71%	100%	7%	42%	15%	32%	3%	1%	***
College or Greater	71%	100%	5%	45%	13%	34%	3%	1%	***
Catholic	81%	100%	3%	48%	12%	34%	2%	1%	***
Protestant	65%	100%	6%	33%	11%	43%	7%	0%	***
Baptist	64%	100%	9%	40%	19%	27%	3%	2%	***
Other	50%	100%	9%	38%	15%	34%	1%	1%	***
18-24	71%	100%	13%	39%	15%	28%	3%	2%	***
25-34	79%	100%	5%	47%	16%	29%	3%	1%	***
35-44	76%	100%	3%	46%	19%	27%	4%	1%	***
45--54	73%	100%	10%	43%	8%	39%	0%	0%	***
55-64	62%	100%	8%	33%	8%	45%	3%	2%	***
65+	50%	100%	0%	36%	13%	47%	2%	2%	***
< than $10,000	68%	100%	7%	30%	17%	41%	4%	0%	***
$10,000/$19,999	57%	100%	13%	39%	13%	31%	1%	4%	***
$20,000/$29,999	73%	100%	5%	42%	19%	30%	4%	1%	***
$30,000/$39,999	79%	100%	3%	48%	10%	34%	3%	0%	***
$40,000/$49,999	76%	100%	6%	42%	13%	34%	4%	1%	***
$50,000/$74,999	71%	100%	11%	47%	13%	28%	0%	2%	***
$75,000 or more	87%	100%	0%	50%	25%	25%	0%	0%	***

The 64 Parishes were divided into 6 geographically defined regions that correspond to the Lottery's regional offices.

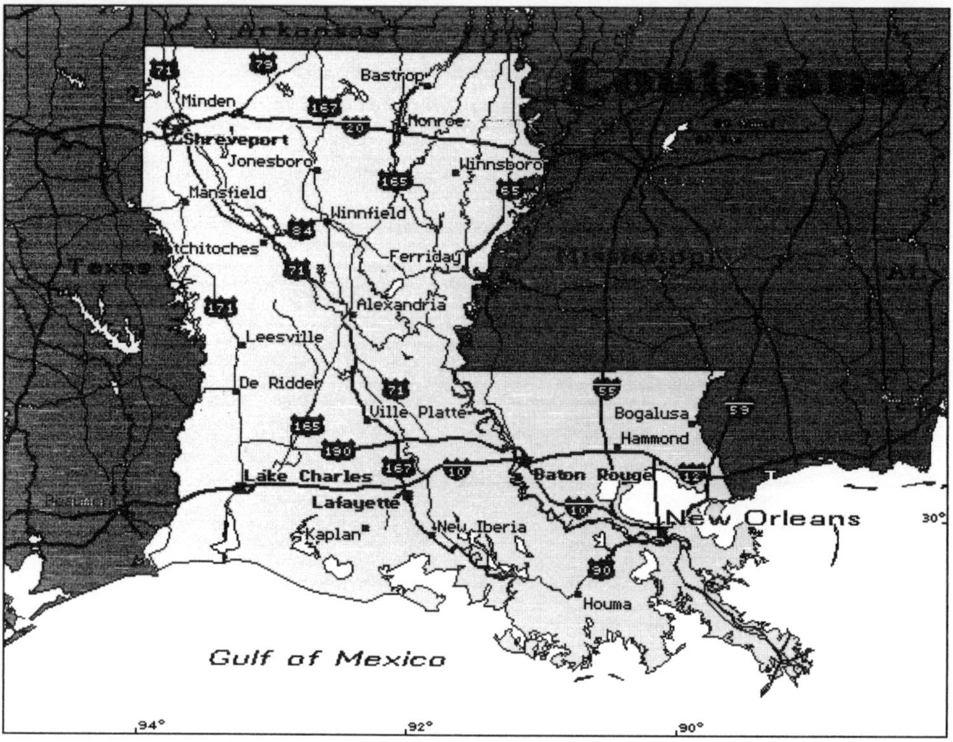

The Lafayette and the New Orleans Regions had the highest percent of those who have "tried" a lottery game to date, about 3%-8% above the state average. This area is the southern part of the State and has the highest percent of Catholics.

The Route 71 corridor extends in a north/south direction through the center part of the state. It encompasses the Baton Rouge, Alexandria and Shreveport Regions. These regions average slightly less than the State average for Lottery game trial. This represents the largest geographic area of the state.

The region with the lowest percent of those who have ever played a Lottery game is Monroe, in the northeastern part of the state. This region is predominantly Baptist and Protestant. The relatively high percentage of current Pick 3 players in this region that otherwise has a low trial rate, is reflective of the high percentage of Blacks, who play Pick 3 at a much higher rate as compared to Whites, living in the City of Monroe.

The following table presents the percent of those who are current lottery players by combination of games currently played:

6. CONCLUSION: BASELINE SURVEYS USAGE TO SUPPORT THE LOTTERY'S MARKETING DECISIONS

6.1. IMPACT TOWARD TERMINAL LOCATION

The baseline research provides the Lottery with the ability to optimize the number of, type of and placement of Lottery retailer locations. Geographic locations that are well suited to serve the Daily Pick 3 player profile vary greatly by business type and location, from a location that primarily serves a weekly or biweekly Scratch player or Lotto player base market. Clustering profiles, such as PRIZM by Claritas can be used as a method of characterizing demographic neighborhoods for all target marketing purposes.

6.2. IMPACT TOWARD PRODUCT MIX

The baseline research also shows the Lottery where opportunities exist to develop targeted marketing programs and to design merchandising campaigns. Periodic surveys at prescribed intervals can track changes in play behavior and evaluate target marketing effectiveness. It is also important to track individual state sales for comparison purposes (4).

6.3. IMPACT TOWARD ADVERTISING BUYS

The baseline survey provides lottery play behavior information on radio listening habits, newspaper subscription and television viewer ship. By applying Lottery play behavior knowledge to the various media the lottery is able to produce more strategic advertising buys.

7. REFERENCES

1. Kopel, P. "Louisiana Lottery Baseline Survey", Report, May 1993.

2. Kopel, P. "Louisiana Lottery Baseline Survey", Report, October, 1992.

3. Weiss, Michael M.J., 1988. The Clustering of America. New York: Harper & Row

4. La Fleur, T. 1991. "Compendium of Lottery Statistics". Maryland: TLF Publications.

A NEIGHBORHOOD OUTREACH PROGRAM: DEMOGRAPHIC FACTORS AFFECTING SITE SELECTION AND ADAPTATION

Linda A. Thompson
AHCCCS
Office of the Medical Director

Susan McGraw-Helms
Az. Dept. of Health Services
Office of Women's & Children's Health

1. INTRODUCTION

Selecting and implementing health programs requires detailed assessment. This paper seeks to examine the demographic factors that were considered for site selections for a health promotion project in Arizona. The processes of selecting health program models and sites where the models will be applied can occur separately or simultaneously. In this case the model was chosen first, and then locations meeting select criteria were identified. This paper does not deal with the effectiveness of the model program, or the criteria that led to the selection of the program, except were the model is relevant to the characteristics of the site.

The goal of the program is to increase the utilization of prenatal care services. The Institute of Medicine has categorized programs designed to improve utilization of prenatal care services into five basic groups depending on the major emphasis of program activities. Two out of the five categories include the majority of activities comprising what is often termed "outreach." Case-finding accomplished through "hot lines" and community canvassing, cross agency referrals and the provision of incentives and social support are key outreach activities. (5)

An outreach approach used commonly in third world nations, and now gaining popularity in the United States, uses peer health workers to deliver health education and selected services to disadvantaged populations. (13) The variety of position titles used to refer to peer health workers includes Peer Health Educators, Lay Health Workers, Community Health Representatives, and Neighborhood Outreach Workers.

Appropriate peer health educators reflect the ethnic, linguistic, cultural, and socio-economic characteristics of the communities they serve. The worker does not function as, and is not viewed as, an arm of an agency. The lay health worker's image is one of an internal resource; a supportive, informed member of the community. By adapting the approach and information to the viewpoint of the recipient, the message is likely conveyed more effectively.

2. DESCRIPTION

UN COMIENZO SANO/HEALTH START is a Lay Health Worker Outreach Program in Arizona. It is designed to address barriers to prenatal care from the perspective of the women and families at greatest risk of poor birth outcomes. Lay health workers reflecting the characteristics of the

---

[1] Identifying women in need of prenatal care.

community promote prenatal care and primary health care for children among low-income pregnant women and their families at the neighborhood level. As neighbors, the lay health workers have entree into homes and are trusted with information that enables them to understand a pregnant woman's needs within the appropriate social and cultural context, allowing them to make applicable referrals and involve other members of the woman's family and support system. (7)

The primary responsibilities of the lay health worker are (6):

(1) To identify pregnant women in their neighborhoods,

(2) To convince the women identified of the importance of prenatal care,

(3) To provide basic prenatal education through home visits and classes,

(4) To serve as a guide through the AHCCCS application process, acting as an advocate when necessary,

(5) To connect women with community resources to fill their other social service needs,

(6) To promote primary health care for children,

(7) To provide personal support,

UN COMIENZO SANO/HEALTH START currently functions in five sites in Arizona. A "site" is a contractor responsible for carrying out the activities of the program. Each site functions in more than one geographical neighborhood.

3. HISTORY

In 1987, the Arizona Department of Health Services, Office of Women and Children's Health, (ADHS, OWCH) and the Rural Health Office of the University of Arizona both began prenatal care outreach / education demonstration projects based on the philosophy that the people from within a neighborhood know best how to reach and teach the others who live in that neighborhood. (7) During the first five years of planning and implementation, ADHS and the Rural Health Office became aware of each other's ventures and combined efforts. These independent demonstration projects have been completely integrated into the Department of Health Services and continue to operate. Based on the experience of the two original projects, DHS expanded the program to additional sites. Following site expansion, a grant was received for expanding the scope of services, and promotion of primary care for children was incorporated into the activities of the lay health workers.

4. METHODOLOGY

The methodology for site selection varies with the evolution of implementation in Arizona. The original sites were selected circumstantially. ADHS had previously run a program referred to as the Perinatal Care Program. This program served economically disadvantaged women without private insurance who did not meet Medicaid eligibility criteria. When mandates expanding eligibility criteria were implemented, funding for this program was transferred to the Medicaid program, assuming

---

2
AHCCS is an acronym for Arizona's alternative Medicaid program. The program is called Arizona Health Care Cost Containment System.

3
The National Association for the Education of Young Children awarded grant money through its Goal One project. The objective of this project is to ensure that no child is "disadvantaged" when beginning school.

that the women would now be likely to qualify. ADHS intended to extend grants to county health departments which would replace the Perinatal Care Program with the UN COMIENZO SANO/HEALTH START model. While four of the fifteen counties attempted to implement services, only one program was managed effectively. This was the Pinal County program in Eloy.

The Southwest Border Rural Health Research Office of the University of Arizona College of Medicine was, in the meantime, conducting a health survey of the migrant farm-worker families in several small communities surrounding Yuma. Though the focus of this survey had to do with tuberculosis, the surveyors inadvertently became aware of the inadequate prenatal care among the population. The Rural Health Office launched a peer educator prenatal program at the same time that the Yuma County Health Department had the opportunity to participate in the ADHS grant. This is how the two entities became aware of each other's similar ventures.

Following the original two demonstration projects, sites were selected by a Request for Proposal (RFP) process. The RFP specified the nature of services, the model to be used, and the preferred target populations. The populations to be served were "Pregnant women at high risk of receiving inadequate or no prenatal care (low income, minority, and teen). African American, Native American, and migrant women are of particular concern." (1) Respondents used their RFP responses to demonstrate why their site was an appropriate selection for implementing UN COMIENZO SANO.

Specific demographic risk factors that have been associated with insufficient prenatal care constituted the selection criteria and were combined with qualitative data to document the need of each of the three successful respondents. Social and cultural features may categorize an area as high risk even where there is not a geographical barrier to care. These traits may include minority status, age, education, birth order, marital status and poverty. Geographical accessibility is also clearly documented to impact prenatal care. (2, 5, 11)

Geographic accessibility is of significance in Arizona where large expanses of land, coupled with low population density, make the delivery of prenatal care services challenging. (2) Federally designated Health Manpower Shortage Areas (HMSA) have been defined and identified according to the following three designations: geographic/service area; population; and facility. When an area is designated as a HMSA, the area has unusually high service needs which are not, or cannot, be met by existing health service delivery systems. (4)

Another designation carries similar implications with regard to accessibility. Federally designated Medically Underserved Areas (MUA) are determined by evaluating several criteria including: The percentage of population 65 years or older; the infant mortality rate over a five year period; the physician-to-population ratio; and the population's inability to access existing services. A population's needs based on ethnic, socio-economic or migratory nature of occupation, together with geographic and/or economic barriers to care, are considered in determining access to services. (4)

Successful respondents to the RFP used a variety of outcome and utilization measures to demonstrate need in their areas. One such measurement is the Kessner Index, a ranking system used to classify the amount and adequacy of prenatal care based on the number of visits made, combined with the stage of pregnancy when care was initiated. (2) A small area analysis measuring the adequacy of prenatal care by the Kessner Index is referenced by the RFP respondents. The incidence of low birth weight, very low birth weight, premature births, and neonatal mortality are also commonly used birth outcome measures. (2, 5, 11) Percentages of these outcomes in a given population are compared with the population as a whole in order to validate the need for prenatal intervention projects.

5. SITE SELECTION ANALYSIS

Analysis of each site according to the criteria described in the methodology demonstrates the appropriateness of the sites selected. This is true even of the original two sites, although these locations were not evaluated by the same process as those selected through the Requests for Proposal.

Each of the five UN COMIENZO SANO/HEALTH START sites is clearly able to demonstrate geographical inaccessibility to care. Each program site is in a community identified as a HMSA as well as a MUA. A recent report indicates that the number of obstetrical care providers in rural areas decreased by 60% in the last 10 years. Indian Health Services has lost 25 obstetric providers in 1989 alone. (12) One program site is located on the Hopi Indian Reservation. A small area analysis, using the Kessner Index as a measurement tool, identified inadequate prenatal care in three of the program sites.

Lack of transportation is another important barrier in the underserved areas. Where health care providers are few and far between, adequate transportation is necessary for women to reach providers. Hopi Health Start is illustrative of some of the transportation issues. The population on the Reservation lives in poverty. Many do not own their own means of transportation. Indian Health Services has adopted the goal that no Native American living on the Reservation should have to travel more than 30 minutes for medical care. (2) Additionally, the Hopi Tribal Department operates a community transit system that is free to all members of the Hopi Tribe who reside on the Reservation. (10) Though this resource is available, many are unable to access the transit system because they have no phone to arrange for service.

Transportation affects the migrant farm worker populations as well. Three of the program sites directly serve migrant farm-worker families. Women served at Clinica Adelante often travel over an hour to reach the clinic. It is not uncommon for them to arrange for transportation through AHCCCS, only to be transported to their appointment and left at the clinic without a ride home. If a woman has had this experience, she may opt to decline her doctor appointments and instead wait for the once-a-month mobile unit to come to her area. (12)

The inner city UN COMIENZO SANO/HEALTH START program, Mountain Park Health Center in South Phoenix, has transportation problems too. The Medicaid managed care plans are mandated to provide medically necessary transportation, but are not instructed as to how to provide such transportation. The most common method is to provide passes for the city bus system. This provision breaks down in the summer, as temperatures can commonly be 110 degrees. A pregnant woman is unlikely to stand waiting for the bus in these temperature extremes. She is even less likely to wait if she will need to bring along her children.

While lay health workers have developed close working relationships with the health care and social service providers in their areas, and have helped women overcome as much of the transportation problem as possible, they often are the only "prenatal care" and/or "social service" the women in their neighborhoods can manage. (6)

The socio-demographic risk factors associated with inadequate prenatal care are evident at the program sites. Each of the communities served by UN COMIENZO SANO/HEALTH START serves an at-risk population cluster where 50% to 70% of the population qualifies for Medicaid. (14)

Statistically significant differences in the utilization of prenatal care between races suggest other social and cultural barriers to care. Black, Hispanic, and Native American women are two to four times more likely than Anglos to receive no prenatal care. Blacks, Native Americans, and Hispanics have the highest infant mortality rates in Arizona, and the low birth weight among blacks is twice that of any other group. (9)

---

[4] Indian Health Services is the entity which, by United States Treaty, provides the comprehensive health services to American Indians living on or near Reservations.

[5] AHCCCS managed care plans are required to provide medically necessary transportation.

[6] Clinica Adelante operates a mobile unit providing health care to the migrant farm-worker families.

6. THE FIVE SITES

Eloy: The town of Eloy grew up equidistant between Coolidge and Casa Grande because African Americans and Mexican Americans were not allowed in either town after curfew. 60% of the population is Hispanic and 73% of the births in Eloy in 1991 were to Hispanic mothers, while only 36.2% of the births countywide were to Hispanic mothers. The town has a higher percentage of African American mothers as compared to the State. A total of 9.3% of the births in Eloy in 1991 were to African American mothers as compared to 4.5% of the births to African American mothers countywide. The area is supported by a lagging agricultural industry. (6)

Yuma: Somerton, San Luis, Wellton The percentage of the population that was Hispanic in 1991 in the whole of Yuma County was 64%. The percentage of the population that was Hispanic that same year was 89.7% in Somerton and 98.7% in San Luis. The communities of Somerton, San Luis, and Wellton have a high percentage of migrant families, far in excess of other areas of Yuma County. The women in eastern Yuma County, Roll, Hyder, and Dateland are part of the Wellton "neighborhood". Though distances are great, they share a common high school, and Wellton is their local shopping area. This area borders Mexico. Women frequently arrive in the county when they are well into their pregnancies and have difficulty connecting with services. (8)

Many of these families fear that their applications for permanent residency may be jeopardized if they access public benefits. Those who have not applied for residency fear being deported. Data available on the socio-demographic characteristics of program participants in Yuma reveals a majority with less than a high school education, approximately 18% who are unmarried, and over 20% who are teenagers. (13)

Clinica Adelante: Aguila, Gila Bend, El Mirage, Queen Creek All of these communities in rural Maricopa County have a high concentration of seasonal/migrant farm-workers. Farm workers tend to be crop specific. They migrate as the crops are ready for harvest in different areas. The workers stay for 3 to 4 months and then follow crops to another area. This prohibits establishing a medical home and conditions requiring follow-up may be neglected. Many of the farm workers are families. (12) While the percentages of births to Hispanic mothers for the whole of Maricopa County in 1991 was 27.8%, the percentage in Aguila was 87%, in Gila Bend 49.9%, in Queen Creek 57.1%, and in El Mirage 86.9%. (8)

Mountain Park Health Center: Keys Community, South Phoenix This program site targets the African American community concentrated in the area between 32nd Street and 24th Street, and Roeser and Wood Streets. The focal point in this community is the Keys Community Center. The Center was formerly a location for drug dealing and violence. Those living in the area, tired of threats to their families, took over the market, gutted the building, built the center, and have reclaimed their community. (8)

The service area of Mountain Park Health Center is characterized as the most culturally mixed in Arizona. African Americans makeup 21% of the population, Hispanics account for 40% of the residents, and Anglos comprise 38%. The area is characteristic of many urban inner-city populations. The largest Medicaid population in the State resides in this area, with 41% living under 200% of the federal poverty line. Teen mothers account for 22.6% of the births, as compared to the average county rate of 12.8%. There are comparatively high infant mortality rates and a high incidence of substance abuse. (2,10) Additionally, 28.3% of women have a low education level as compared to 14.2% for the county as a whole.

Hopi Health Start: First Mesa, Second Mesa, Third Mesa, Moenkopi Each of the mesas includes three to four traditional Hopi Villages. All four of the primary program sites serve Hopi residents; the village of Moenkopi also includes some Navajo residents. Although this population group has a distinct health care delivery system, it exhibits the same kinds of cultural and social barriers to prenatal care that are found at the other sites. (6)

Public health representatives of the Hopi Nation state that the number of teenagers becoming pregnant is increasing, however the increase is not quantified in this source. (10)

---

7 See the discussion in this paper regarding the program's definition of "neighborhood."

7. DEFINING A "NEIGHBORHOOD"

All sites used units or boundaries different from one another in defining their service areas. Mountain Park chose census tracks to quantify the minority population, identify the area-specific poverty, and define the "neighborhoods" served. They also took data from the clinic's patient profile to further demonstrate demographic risk factors and poor outcomes. (10) There is an implication in this method that the clinic's clientele all come from the four census tracks specified, although this may not necessarily be true.

Hopi Health Start looked at the births in specific zip codes, yet simultaneously acknowledged this was problematic because zip codes, county lines, and Reservation boundaries all overlapped. Information from the Bureau of Indian Affairs was used to forecast the number of women who would be served in the specific Hopi neighborhoods. (10)

Clinica Adelante relied on a study that identified migrant farm-worker families by identifying the number of acres of labor intensive crops in the state, and then breaking the information down by county. Target areas within the clinics service area were then selected.

Defining the "neighborhood" to be served became even more complex during implementation. Several concepts needed to be worked through in determining who the lay health workers would visit. The issues included social, geographical and safety aspects of neighborhood canvassing. Questions were used to spark discussion and provide a comfortable framework within which the lay health workers could begin to work.

- Would they make "cold calls," door-to-door at the homes of people they didn't know?
- How far from home would they travel?
- How far would you let your child go to play before it gets dark?
- As a woman, how far from your house would you walk by yourself after dark?
- How far could you walk from your house and pass someone working in their front yard and smile and wave, even if you didn't know their name?

This process led to the realization that there were some areas in the state where the model would never work. In areas where people don't identify with neighbors or trust their neighbors, visits from a friendly neighbor were not accepted. This is a critical consideration in selecting sites.

The program model and the RFP, required that the lay health workers live in the neighborhoods that they served. Did this mean that if one of the lay health worker's clients asked the worker to visit her pregnant cousin five miles away, she would have to say no?

A conceptual definition of "neighborhood," evolved. The idea of a geographic neighborhood was too limiting. Lay Health Workers do not make cold calls, they visit people they "know." They do serve their own neighborhoods, but they also include their own personal network (1,9,10), which may take them outside their immediate area.

This concept is beginning to evolve even further. There are areas in Arizona that are very remote, more frontier in nature than rural. There is a lay health worker program that has been serving this area and will soon become part of UN COMIENZO SANO/HEALTH START. In this "neighborhood," neighbors may live 25 miles apart. The area has a supportive network in place and is able to use the model.

8. CONCLUSION

Selecting areas for program implementation is a complicated process involving socio-demographic and geographic considerations in terms of the program model and the target population. Matching models to geographic and socio-demographic qualities are part of the process. Once sites are selected, flexibility in interpreting the model is necessary, as the concept of neighborhood illustrates.

UN COMIENZO SANO/HEALTH START sites selection involved several interacting considerations. The next step to truly test the appropriateness of the site selection criteria would be a complete and thorough evaluation, measuring the numbers served, incidence of true case-finding, changes in utilization of prenatal care, birth outcomes and establishment of primary care for children. Thorough and complete program data is necessary to analyze the effectiveness of each.

Compiling program data has been problematic, partially because of the nature as a characteristic of the model chosen and partially due to specific site locations. Data collection is inconsistent with the role of the lay health worker. The model emphasizes neighbors helping neighbors, yet when a neighbor comes to visit, they do not generally take notes. Lay health workers were not selected for their record keeping or data collection skills, but rather for their standing in their communities. This issue of inconsistency is common to each site.

There are individual site problems with data related to turnover. At some sites there has been a rapid turnover in lay health workers and program coordinators. Prior to expanding the scope of services to include primary care for children, data were collected throughout the pregnancy and turned in after the postpartum visit. This meant that if a lay health worker visited a woman two or three times and left the program before the client's baby was born, there would be no data. Now that the scope of services extends to two years after the conclusion of the pregnancy, this issue is further complicated.

In spite of these issues, data has been collected. The basic information needed for future direction exists. ADHS is presently seeking expertise to analyze the compiled data and the collection process. Reconciliation with secondary data sources, such as birth certificates and Medicaid eligibility records will be part of the process. A future requirement for determining the appropriateness of the model and the selected sites, as well as for ensuring program survival and growth, is to resolve the issues surrounding data collection and ultimately evaluation.

9. REFERENCES

1. Clinica Adelante, Inc., 1991. "Perinatal Outreach Services." Response to Arizona State Department of Health Services Request for Proposal no. 2-228.

2. Gallagher, K., 1989. "Use of Prenatal Services Special Populations Needs Assessment." Arizona Department of Health Services Office of Planning and Health Status Monitoring, Unpub. Report.

3. Gersten, Joanne C., and Mrela, C., 1991. Closing the Decade: Arizona Health Status and Vital Statistics, 1980-1989. Arizona Department of Health Services Office of Planning and Health Status Monitoring

4. Gordon, R.J., 1987. Arizona Rural Health Provider Atlas, Rural Health Office Department of Family and Community Medicine, University of Arizona pp 141-147.

5. Institute of Medicine, 1988. Prenatal Care: Reaching Mothers, Reaching Infants, Washington, D.C.: National Academy Press.

6. McGraw-Helms, S., 1992. "Un Comienzo Sano/Health Start Annual Report" Arizona Department of Health Services Office of Women's and Children's Health. Unpub. Report.

7. McGraw-Helms, S., 1991. "Un Comienzo Sano/Health Start Annual Report" Arizona Department of Health Services Office of Women's and Children's Health. Unpub. Report.

8. McGraw-Helms, S., 1993. "Semi-Annual Report for Goal One Un Comienzo Sano Health Start Arizona Department of Health Services Office of Women's and Children's Health. Unpub. Report.

9. Memorial Family Health Center, 1991. "Funding Request for Perinatal Outreach Program to African American Women (Un Comienzo Sano). Response to Arizona State Department of Health Services Request for Proposal no. 2-228.

10. Northern Arizona Area Health Education Center, Inc., 1991. "Hopi Health Start: Perinatal Outreach Services." Response to Arizona State Department of Health Services Request for Proposal no. 2-228.

11. Study Committee on Services to Pregnant Women, 1992. Final Report Presented to Fife Symington, Governor of the State of Arizona, Senator Peter Rios, President of the Arizona State Senate and Representative Jane Dee Hull, Speaker of the Arizona State House of Representatives.

12. Trotter, R.T. and Langdon, M. 1989. "Arizona Migrant and Seasonal Farmworker High Density Areas." Department of Anthropology Northern Arizona University Unpub. Report.

13. Warrick, L.H., Wood, A.H., Meister, J.S., de Zapien, J.G., 1992. "Evaluation of a Peer Health Worker Prenatal Outreach and Education Program for Hispanic Farmworker Families." Journal of Community Health 17(1):3-26.

14. Un Comienzo Sano/Health Start program data 1992, obtained from Clinica Adelante, Northern Arizona Area Health Education Center, Pinal County, and Yuma County through the Arizona State Department of Health Services.

SITE ANALYSIS, DECISION MAKING AND THE ANALYTICAL HIERARCHY PROCESS

A. William Dakan
Department of Geography and Geosciences
University of Louisville
Louisville, Kentucky 40292

1. INTRODUCTION

Analytical methods used in site selection such as locational-allocational programming, gravity models, regression techniques and the increasing use of business-level geographic information systems give the discipline of site selection a patina of authoritativeness. I know, as do many site analysts, working as either in-house analysts, academicians or private consultants, that clients do not always accept either our methods or conclusions as axiomatic truths. Decisions about the ultimate location of an intended facility are presumably made on the recommendations of the site analyst and are subsequent to the completion of the analysis. It is not unlikely that a decision maker may selectively use certain observations of the analyst or may substitute his/her individual experience or perception of the market area for the analytical impeccability of the consultant's report. Decision makers may not engage in any formal analysis at all, creating a "site analysis" team of four corporate vice-presidents, a fast car and a clean windshield.

This paper presents a method of incorporating the market vision and perceptions of the decision-maker or client into the analytical framework of the site analyst as an antecedent step to the analysis. In this specific instance, a procedure known as the Analytical Hierarchy Process (AHP) was used retrospectively to determine the "correctness" of decisions made by an entrepreneur in using multiple criteria to evaluate and select alternate sites for an expanding enterprise. This retrospection came about in an unusual way. I was asked by the Louisville Entrepreneur Society, a group of actual and would-be small business owners, to speak to their group. I shared the podium with a Market Research Analyst in private practice and a young businessman who ran three very successful franchised bakeries which specialize in natural and whole grain breads, muffins and cookies. The perspective each presenter was to bring to the session was determined by the program chairman. The market analyst was to offer advice on choosing a consultant and what to expect from such a consultant. The baker was presented as a maverick who prospered in spite of formal analysis of potential sites. I was supposed to be the ivory tower theoretician, not an enviable role among potential beauty shop operators, nascent realtors and budding franchisees. In the newsletter preceding the meeting, I was identified as "surprisingly, a geographer". As a result of our presentation it was decided that I would evaluate the baker's choice of two additional sites and show how similar results could be obtained in a formal analysis. AHP was used to (1) determine the weights the baker assigned to various site factors and (2) gauge the extent candidate sites reflected his priorities.

2. THE ANALYTICAL HIERARCHICAL PROCESS

The AHP begins with defining and prioritizing desirable (or undesirable) site characteristics. These characteristics can be hierarchical in that decision criteria may exist at different levels of abstraction. For example, "accessibility" may be a high level abstraction made more concrete by substrata consisting of such items as parking, visibility from the street, traffic signaling and other measurable characteristics. The software used in this paper (5), discussed below, permits only a single level of specificity in criteria. More complex models can be constructed. A complex model for developing an energy strategy for China uses distinct time periods in future energy resource development as the top level in the hierarchy.

Once the decision criteria are elicited and defined, a pair-wise comparison of each criterion with each of the remaining criteria is made. The relative importance of each criterion is weighed against the overall decision objective. All possible pairs of criteria, including those which seem to be mutually exclusive or independent, must be compared. The comparison is made by applying the 1 to 9 scale noted in Table 1.

TABLE 1
AHP Measurement Scales

Value	Definition	Explanation
1	Equal importance	Two criteria contribute equally to the decision objective
3	Moderate importance of one over another	Experience and judgement favor one criterion over the other
5	Essential or strong importance	Experienceand judgment strongly favor one criterion over the other
7	Demonstrated importance	A criterion is strongly favored and its dominance is demonstrated in practice
9	Extreme importance	The evidence favoring one criterion over the other is of the highest possible order of affirmation
2,4,6,8	Used to compromise between judgement levels	

After Saaty (5) as reproduced in Bowen (1)

The decisions, valuing one criterion over another, are arrayed in a matrix. The elements of this judgement matrix are assigned one of two values, the scaled 1...9 judgement score or the reciprocal of the judgement score. For example, if the element aij is equal to 9, then the element aji is the reciprocal, .111. Major diagonal elements are 1.00.

Criterion weights are calculated for each row of the matrix by finding the geometric mean of the row elements. To do so, the elements are multiplied across all k elements and then the kth root is extracted. These geometric row means are normalized by dividing each row mean by the sum of the row means. (1) An alternate method is to normalize by finding the maximum eigenvalue for the judgement matrix. (2) In either case, the resulting vector consists of the priority weights of each of the criteria, weighed against all of the remaining criteria. These weights are presented as a vector and should sum to unity.

The ultimate decision outcome applies the criteria against specific scenarios, in the present case, optional sites. Again, paired comparisons of each site, for each criterion, are created using the same 1...9 scale to judge the criteria. Matrices of site by site comparisons, one matrix for each criterion, are created and normalized using the same procedure to create the judgement matrix. The normalized weights are the relative rankings of the sites on each of the performance criterion.

The final computational step is to calculate a single priority weight for each site, resulting in the possible selection of the "best" site, given their relative performances on the stated criteria. The priority of site j is the sum of the product of criteria weights, wi, and qij, the priority weight for site j with respect to criterion i.

Because the AHP is used to clarify values and to make judgements, it is possible that there may be a problem with consistency within the judgement and comparison matrices. It is assumed that the judgements are transitive. That is, if criterion S is preferred over T, and T over Q, then S is preferred over Q. Inconsistencies can also arise in evaluating alternatives. The most serious consequence of an inconsistent judgement array is rank reversal. A peculiar effect can occur when an new alternative is introduced. (2) For example, if an automobile purchaser compares a Lexus, an Infiniti, a Cadillac and a Mercedes, and then a BMW is introduced to the decision matrix, rank reversals among the unrelated Lexus and Cadillac can result. Ranks involving the close neighbor to the BMW, the Mercedes, are even more likely. The problem of correcting rank reversals is covered in Golden. (4)

The AHP produces a consistency index to evaluate the internal consistency of the decision maker. The consistency index is a decimal fraction resulting from comparing the observed matrix of decision weights with a matrix of "perfect consistency." (1) Such a matrix could exist if the relative weights among cases were the result of the measurement of physical characteristics such as distance, land value or sales per square foot. As a general rule, consistency indices greater than 0.1 indicate that the judgements are too inconsistent to be reliable. Since each decision matrix has a corresponding consistency index, the offending judgements can be reconsidered without totally reconstructing the entire analysis.

3. A SITE ANALYSIS APPLICATION

Site analysis should be a key application area for AHP. Accessibility to the procedure has been made easier by a spreadsheet model developed by Bowen and made available in conjunction with other spreadsheet models in urban and regional analysis. (1) A three-level hierarchical model has been used to locate an electronics plant in Japan. (7) In this model, the highest level of abstraction are "technical, economic and social" factors. Under "technical" considerations are items such as weather. At the operational level of the model, weather elements include rain, snowfall, earthquakes, volcanic eruption and potential injury from salt. Initially working without a framework such as the AHP, the site analysis team (mostly technical and administrative personnel from within the company) found it difficult to reach a consensus on any particular site. The candidate sites were rated excellent, fair, less than fair and poor on over 120 specific items. Using the AHP clarified the issues and resulted in a site that was initially ranked below a competing site using a simple check list of the same variables considered with the AHP.

The most immediate example is a retrospective analysis of the decision to expand by a local Louisville baker. His bakery specializes in "natural" breads, and offers a limited line of cookies and muffins. As a very young entrepreneur, he engaged in a time-consuming and totally observational procedure to acquire the base site where all of the baking is done. He sat outside several possible sites for weeks to determine which had the highest traffic and the best parking. In addition he used a general familiarity with Louisville to locate within a middle-class neighborhood. The enterprise appears to be a success and he has established two satellite sites. These sites were selected by the owner after a brief period when he collected the zip codes of customers at the original establishment. He then reapplied his concern for traffic and parking to the appraisal of the sites. This analysis is the result of recasting his thinking in a format that can be used in the AHP and applying the technique to his two new sites as well as three additional real, but unconsidered, sites to determine if his locational decision could be supported. The factors, and the priority values from the AHP are:

(1) Population (POP) within the general area of the site. Priority value is .255, second highest.

(2) Traffic (TRAF) in the general area of the site. The priority value is the highest at .448.

(3) Income (INC) based on impressions of housing in the area and a generally held belief as to the identification of "better neighborhoods". The priority is a relative low .064.

(4) Attractiveness (ATT) of the building or space to hold the site as viewed by the potential tenant. Attractiveness is a low .033 in priority.

(5) Parking (PARK) as actually measured by the number of parking spaces available to the site. Parking is third, with a value of .132.

(6) Distance (DIST) from the existing facility, where the product is made. Distance ranks slightly higher than income with a value of .068.

The "client" did a rough pairwise comparison of the desired attributes and kept strictly to the 1 to 9 scale proscribed by the AHP. The results, in AHP format are listed in Table 2.

TABLE 2
Attributes Weighted For Analysis By AHP

	POP	TRAF	INC	ATT	PARK	DIST
POP	1.000	0.333	5.000	7.000	3.000	7.000
TRAF	3.000	1.000	7.000	9.000	7.000	5.000
INC	0.200	0.143	1.000	5.000	0.333	0.333
ATT	0.143	0.111	0.200	1.000	0.143	1.000
PARK	0.333	0.143	3.000	7.000	1.000	3.000
DIST	0.143	0.200	3.000	1.000	0.333	1.000

Source: the author.

The consistency index for this matrix is .170 which is higher than should be allowed for the analysis to continue. In more critical situations, the client or decision making group would be requested to re-evaluate the attributes until the inconsistencies were reduced to acceptable levels.

The next step is to apply the attributes to a set of sites. The base site for the bakery is in a middle-class suburb of Louisville on the generally more affluent Eastside. The site was used in various types of food service so that substantial reconstruction was not needed for conversion to a bakery. The site is a side street in a commercial district located adjacent to Shelbyville Road, a major east-west arterial. The other sites, and possible candidate include:

(1) Middletown - (MIDT) this site is farther out Shelbyville Road from the existing bakery and was the second satellite to open. The population is similar in density and character to the original, traffic is substantial, the site is moderately attractive, the parking is considerable and the site is relatively far from the original.

(2) Bardstown Road - (BARD) this site is part of a new building on a traditional shopping street that has undergone a substantial transition from a streetcar suburb to a partially developed boutique and restaurant row. The site is part of a small complex which includes a record store and a video chain store. Traffic is high, parking is marginal and the building is attractive according to the client. The site is within four miles of the original bakery, a plus since all baked goods must be delivered.

(3) Downtown - (DOWNT) the appropriate site is within the highest concentration of daytime population in the concentration of building around the Downtown Shopping Mall known as the Galleria. There is a number of attractive sites. It is difficult to apply attribute priorities designed to evaluate suburban and free-standing sites to typical central sites.

(4) Hikes Point - (HIKEP) another suburban site is about five miles farther out and off of Bardstown Road. The income drops somewhat, the traffic reaches congestion levels during the day, parking is adequate and the site environment is especially unattractive.

(5) Prospect - (PROS) this site is the farthest from the bakery and is located in the most affluent, but relatively sparsely populated, area of the county. The principal thoroughfare is US 42 and the traffic is moderate, but increasing as development extends rapidly in the area. The site is attractive as might be expected in upscale developments.

The evaluation of each site against the attributes is presented in Table 3.

TABLE 3
Site Evaluation

Population

	MIDT	BARD	DOWNT	HIKEP	PROS
MIDT	1.000	0.333	7.000	3.000	3.000
BARD	3.000	1.000	9.000	3.000	3.000
DOWNT	0.143	0.111	1.000	0.333	0.200
HIKEP	0.333	0.333	3.000	1.000	3.000
PROS	0.333	0.333	5.000	0.333	1.000

Traffic

	MIDT	BARD	DOWNT	HIKEP	PROS
MIDT	1.000	1.000	5.000	3.000	7.000

BARD	1.000	1.000	7.000	5.000	7.000
DOWNT	0.200	0.143	1.000	0.200	0.200
HIKEP	0.333	0.200	5.000	1.000	2.000
PROS	0.143	0.143	5.000	0.500	1.000

Income

	MIDT	BARD	DOWNT	HIKEP	PROS
MIDT	1.000	3.000	7.000	3.000	0.333
BARD	0.333	1.000	5.000	3.000	0.143
DOWNT	0.143	0.200	1.000	0.200	0.111
HIKEP	0.333	0.333	5.000	1.000	0.200
PROS	3.000	7.000	9.000	5.000	1.000

Attractiveness

	MIDT	BARD	DOWNT	HIKEP	PROS
MIDT	1.000	0.333	0.200	5.000	5.000
BARD	3.000	1.000	0.143	7.000	1.000
DOWNT	5.000	7.000	1.000	9.000	7.000
HIKEP	0.200	0.143	0.111	1.000	0.143
PROS	0.200	1.000	0.143	7.000	1.000

Parking

	MIDT	BARD	DOWNT	HIKEP	PROS
MIDT	1.000	3.000	7.000	3.000	0.333
BARD	0.333	1.000	5.000	0.333	0.333
DOWNT	0.143	0.200	1.000	0.333	0.143
HIKEP	0.333	3.000	3.000	1.000	0.333
PROS	3.000	3.000	7.000	3.000	1.000

Distance

	MIDT	BARD	DOWNT	HIKEP	PROS
MIDT	1.000	0.200	0.200	0.333	2.000
BARD	5.000	1.000	3.000	5.000	7.000
DOWNT	5.000	0.333	1.000	3.000	7.000
HIKEP	3.000	0.200	0.333	1.000	3.000
PROS	0.500	0.143	0.143	0.333	1.000

When all of the factors are known, the final prioritization of the sites can be calculated. The results are listed in Table 4, for the priority on each .

TABLE 4
Site Evaluation Priorities By Attribute

	POP	TRAF	INC	ATT	PARK	DIST
Middletown	0.266	0.342	0.232	0.159	0.271	0.071
Bardstown	0.429	0.401	0.133	0.165	0.117	0.477
Downtown	0.037	0.042	0.033	0.540	0.041	0.279
Hikes Point	0.154	0.128	0.095	0.030	0.156	0.128
Prospect	0.114	0.088	0.507	0.106	0.416	0.045
Consistency	0.095	0.114	0.104	0.249	0.105	0.010

In the above table, consistency ratings are relatively low for a first pass at the data. The most elusive attribute, attractiveness, has the most inter-site inconsistency. Conversely, the attribute that is the most precise because it is totally measurable, distance, is the most consistent. Overall, the composite site priorities are

TABLE 5
Site Priorities

Alternative	Priority
MIDT	0.282
BARD	0.351
DOWNT	0.073
HIKEP	0.133
PROS	0.162

It is clear that, in spite of considerable consistency problems, which could be corrected, the two active sites exceed, substantially, the utility to the client of the three hypothetical sites. Working with some rudimentary knowledge of optional sites, and in an informal decision-making context, the client made choices which are justified by the AHP.

3. CONCLUSIONS

In spite of a clear utility of the AHP to the tasks of site analysis and urban planning, the procedure is not common in these fields. The technique seems most appropriate in those cases where decision making involves multiple, and often conflicting goals. The technique seems particularly useful where multiple decision makers (real estate, marketing, production and human resources staffs in site analysis or environmentalists, builders, transit officials and neighborhood leaders in planning) need to resolve conflicts. (6) When multiple groups are involved, each with a particular agenda, the consistency index can be viewed as also the level of conflict within group goals-setting. The procedure is particular adept at combining the expert opinion of experienced clients , who are still naive about the availability of data and techniques, with the analytical capability of locational specialists, who, in turn, are not always well versed in the salient requirements of specialized businesses.

4. REFERENCES

1. Bowen, W.H. 1993. "AHP: Multiple Criteria Evaluation," In R. Klosterman, R. Brail and E. Brossard, Spreadheet Models for Urban and Regional Analysis. New Brunswick: Rutgers Center for Urban Policy Research. 333-356.

2. Dyer, J. 1990. "Remarks on the Analytic Hierarchy Process." Management Science, 36: 249-257.

3. Gholamnezhad, A.H., and S.R. Xia, 1984. "Formulating Energy Strategies and Policies for China." Environment and Planning B, 11: 213-228.

4. Golden, B.L., E.A. Wasil and P.T. Parker (eds), 1989. The Analytic Hierarchy Process. New York: Springer-Verlag.

5. Saaty, T. 1980. The Analytic Hierarcy Process. New York: McGraw-Hill.

6. Saaty, T. 1989. "Group Decision-making and the AHP". In B.L. Golden, E. A. Watsil and P.T. Parker (EDS), The Analytical Hierarchy Process, New York: Springer-Verlag. 59-67.

7. Tone, and Yanagisawa, 1989. "Site Selection for a Large Scale Integrated Circuits Factory". In B. L. Golden, E.A. Wasil and P.T. Parker (EDA), The Analytical Hierarchy Process, New York: Springer-Verlag. 242-250.

COMMUNICATION OF SPATIAL DATA DURING HURRICANE ANDREW

Ute J. Dymon
Department of Geography
Kent State University
Kent, Ohio 44242-0001

I. INTRODUCTION

In 1987, the United Nations adopted a resolution declaring the 1990s the International Decade for Natural Disaster Reduction. (1) The U.S. Senate and House of Representatives endorsed the decade concept in resolutions passed the following year. (2) This has been applauded as an opportunity to employ technology in the service of humans. At a time of extraordinary efforts to respond to ever increasing natural and technological disasters, governments are attempting to locate and define hazards and to prepare for potential emergency events with new technology. Actions and information flows necessary for successful emergency management are based on the knowledge of spatial relationships within the crisis situation which are best presented through the medium of a map. (3) The use of modern technology in support of emergency mapping was investigated immediately after Hurricane Andrew, a Class 4 storm (4), struck southern Florida in August, 1992. Funded by the Natural Hazards Research and Applications Information Center, this research focused on two questions: How were maps used by emergency personnel and by the public during and after Hurricane Andrew? and, What kinds of maps were needed during and after the disaster? Findings came from interviews conducted in the Disaster Field Office (DFO) in Miami with emergency personnel, and from interviews with hurricane victims in the communities of Goulds, Perrine and Homestead.

In keeping with findings from public interviews conducted in previous disaster research, similar public responses were found concerning map use during Hurricane Andrew. The public didn't think about using maps. Moreover, they didn't own any maps to use anyway. They felt maps produced by the media, especially television maps, were the most important to them and kept them well informed.

Generally, there was a lack of available maps. This was even the case with the DFO itself. The formerly empty Eastern Airlines Terminal at Miami International Airport was converted to serve as the Disaster Field Office for nearly 10,000 emergency personnel. This cavernous building became a confusing maze of agency offices, yet no map of the building was made available for people to find their way around in it.

2. THE EARLY RESPONSE SITUATION

What maps did the emergency managers and workers have to work with? Two basic map uses existed at the DFO. Managers needed maps for decision-making, emergency personnel needed them to conduct their field work.

2.1 MAPS DISTRIBUTED BY DADE COUNTY

Already existing maps were made available in the DFO by Dade County. Immediately after the storm, one Dade County official called Rand McNally and ordered all the copies they had in stock of their State of Florida road map. These 5,000 maps were distributed in the DFO and used widely. Field workers expressed their dissatisfaction with these maps because they did not have enough information on them about city landmarks. With an overwhelming number of highway exit signs and street signs missing, information about landmarks was vitally needed for emergency personnel to find locations. In high demand from Dade County's files were population maps showing distributions of ethnic groups and land use maps.

2.2 CRISIS MAPS

Crisis mapping was a major focus of this research. Crisis maps are maps made on the spot during or after a disaster to aid in handling the physical agent or the human response to the disaster.

2.2.1 Handmade Crisis Maps

Most of the federal and state agencies had some sort of crude crisis map on their office walls to facilitate decision-making. These crisis maps were in most cases being built up from a base map such as the Rand McNally map of Florida or from a map brought by a map-oriented member of the agency to the DFO. Often a mylar sheet covered the map, and grease pencils, pins and colored labels were being used to add information. These were stationary crisis maps that stayed on the office walls and could not be copied for field use, but were the centers of attention during the DFO decision-making process.

2.2.2 Dade County GIS Crisis Maps

Dade County had a geographic information system (GIS) in operation in their county office. Immediately after the hurricane struck, this GIS was used to produce two maps that were in high demand: 1) a map of all the public buildings which existed before Hurricane Andrew, and 2) a road network map. An interesting sidelight to this GIS-backed crisis map production was the problems that resulted when a professional technocrat was faced with the task of crisis mapping. One county official revealed that the professional operating the Dade County GIS did not want to release his GIS-produced maps unless they were completely perfected and represented the highest level of cartographic standards. He would not cooperate with the spirit of, and tenets of crisis mapping which are based on the idea that any simple, crude map is a help under the time pressures of disaster response. The County official got the crisis maps he wanted by taking computer-generated maps the GIS operator had discarded out of the wastebasket.

2.2.3 Date County GIS Damage Assessment Map

The Dade County GIS produced a damage assessment map which served an important function in the early response stage after Hurricane Andrew. The map was hand colored on the basis of the aggregation of police reports. Information for this map was collected by the local police then added into the county-wide GIS. Since the police couldn't get into all areas because of the blowdown (such impenetrable areas became severity boundaries) and police helicopters were unavailable since their hangar had collapsed at Tamiami Airport, the police had to make some assumptions in gathering their data. Although in essence subjective, the information came from a usually reliable source, the police on foot. This resulted in the first real damage report developed with systematized reliability and converted into a map. Widely distributed throughout the DFO, this map was a vital technical addition to the work of emergency crews during the early response stage of Andrew.

2.2.4 Aerial Imagery and Photo Interpretation

A critical element in the response effort involved the acquisition and interpretation of aerial imagery for damage assessment. Dade County sought state sources for aerial imagery of the disaster area. Nine tapes of the Florida coast were available from the South Florida Water Management District. The state had some remotely sensed images from the French satellite SPOT and had a Department of Defense trained photo interpreter on their staff.

A representative from the Governor's Liaison Office visited the DFO a few days after the storm. He was making efforts to coordinate the state GIS because he was preparing a paper for Congress on the use of state level GIS. He made available to the County a damage assessment map produced by the state, but the county had to pay $55 to have it sent by express mail. Addressing the issue of county-state coordination, the County official explained: "The state is dealing with different problems than we are. Dade County is only 1/67th of their problem. They have a long-term orientation; they know a disaster can happen in one of any of the other of Florida's counties."

A Dade County official also attempted to secure a set of aerial photographs from the federal government for use in damage assessment. It was known that the military had flown over the area immediately after the hurricane. He tried for three days to get photos from the federal government. He learned that photo work had been done, intelligence people had interpreted the photos, but it would take weeks for him to get permission to access the data and materials at an unclassified level from intelligence agencies. He talked to people in the Pentagon and was sent two photo sets from a company titled Continental Photo Services. However, the County was unable to compare this imagery and achieve photo interpretation for damage assessment because no photo interpreters were available to the County. This whole effort on the part of the County gives some measure of the degree of local-federal coordination in using aerial imagery.

3. MAPPING FOR EMERGENCY RESPONSE IN THE SECOND TWO WEEKS

Two weeks after Hurricane Andrew struck, a historic event took place in the Andrew Disaster Field Office. A private Miami computer firm, Digital Matrix Service, Inc. (DMS), received permission from the Planning and Information Emergency Support Function 5 of FEMA to set up equipment in the DFO to provide emergency mapping from their GIS called InFoCad. Realizing that one of the databases they had recently prepared for a client provided all the street locations from Key West to central Florida, DMS felt they had valuable information already in a form which might help in coordination of emergency response efforts in south Florida. They offered to provide their mapping services free to FEMA for thirty days. The Army helped them move a workstation and a pen plotter into the DFO, and DMS contacted other computer firms around the country with whom they had a working relationship to secure technical support and service for this voluntary endeavor.

3.1 IN-HOUSE GIS MAPPING PROVIDES CRITICAL COMMUNICATION

FEMA solved a critical communication problem by having an in-house mapping capability. Two weeks after Hurricane Andrew, many victims were still stranded by the storm and had no communication with the outside world. The White House Task Force had arranged to have the Goodyear Blimp stationed near the worst hit neighborhoods to furnish messages flashed on the blimp in both Spanish and English to victims. Finding the precise location of the victims was the problem. Literally, as soon as the DMS computers were plugged in at the DFO, an emergency worker requested a map showing the geographic coordinates of all the neighborhoods in Homestead and Florida City. The captain of the blimp used this coordinate map and his on-board global positioning system equipment to find the location of the requested neighborhoods. Information about aid, shelters, field kitchens and hospitals was conveyed to those in need. Word of the successful use of this first custom-made map moved quickly throughout the Disaster Field Office and started a stream of ongoing requests for specially prepared crisis maps needed by emergency managers. The DMS personnel expanded their mapping capability by bringing in more people and more equipment to meet the rapidly increasing demand for on-site mapping.

3.1.1 GIS as an Analytical Tool

The established database employed by DMS covered the entire land area affected by the storm, which in turn facilitated queries and analysis. As many emergency personnel began to comprehend the value of the geographic information system being implemented, requests for analytical maps grew steadily. In one case, GIS operators were asked to identify polling places for the Primary Election which had been postponed because of Andrew. It was not an easy task to select polling sites. The law requires they be a certain distance away from any military activity, but with over 28,000 military personnel deployed throughout the area, this became a challenge. After a few such map challenges were successfully solved by GIS-based map production, requests for more analytical maps were received by DMS. In the end, sixty-three percent of the total number of maps produced by DMS required the InFoCad GIS be applied as an analytical tool.

3.2 THE PROBLEM OF AUTHORITY OVER DATABASES

To further enhance information flow and to educate those providing the 12 Emergency Support Functions in the DFO about the immediate efficacy of GIS applications, the Director of FEMA's Emergency Support System 5, Planning and Information, invited key response agencies to a meeting. The purpose of the meeting was to seek cooperation and understanding about the need to coordinate and integrate all existing available database resources into the GIS being applied daily at the DFO. During the meeting it became clear that the majority of key personnel had little understanding about what a GIS is or what it can do. Official representatives from the major response agencies were asked to provide lists of what data they had available in digital form. It became obvious that the Joint Task Force could provide much of the immediately needed data, but its members were reluctant to share with private agencies any raw data in digital form although they offered support and interpreted data. In contrast, local and state agencies shared their digital data, and many private companies, intent on relieving the suffering of the hurricane victims, shared their resources. The longest discussion during the meeting centered on the issue of the need for a central authority, a "GIS Czar" as it were, to preside and provide priorities for the sharing of, and integration of data from emergency response agencies, other federal agencies, and the Department of Defense.

3.3 NUMBERS AND TYPES OF MAPS USED BY EMERGENCY WORKERS

A daily log of the maps requested between September 6th and October 27th was kept by Digital Matrix Service. In that time period, there were 633 map requests. Three types of map data were requested. TYPE I data, or point data, showing the site locations of a specific data set (see Table 1) were included in 99 percent of the map requests. Another 38 percent of the maps represented requests for the addition of TYPE II data, or aerial data, which was added to the TYPE I map data (see Table 2). From a production perspective, maps that required TYPE III data were more complex and time consuming to create. These maps required scanning, rectifying, and manipulation of data. Eleven percent of the requests were for TYPE III data (Table 3). (5)

Table 1
TYPE I DATA

- Disaster Application Centers
- American Red Cross Kitchens
- American Red Cross Warehouses
- American Red Cross Headquarters
- Metro Dade Burn Sites
- Unemployment Claims Offices
- US Army Kitchens
- Community Health Centers
- MERS/MATTS
- Emergency Reception
- Portable Toilets
- Trailer Parks Centers
- HRS Service Centers
- American Red Cross Service Centers
- Metro Dade Trash Sites
- US Army Tent Shelters
- American Red Cross Hurricane Shelters
- Disasters Medical Assistance Teams
- IRS Service Centers

Table 2
TYPE II DATA

- Building and Zoning Damage
- Mortalities
- Contractor Zones
- Damage Assessments
- Dunn and Brad Street Census Data
- Evacuation Zones
- FIRM Data
- FPL Siren Locations
- FPL Siren Buffers
- Hotel Locations
- Polling Locations
- Voting Precincts
- Surge Contours
- Surge Polygons
- Zip Code Zones
- Hazardous Mitigation

Table 3
TYPE III DATA

- Scanned and rectified aerial photography provided by the Army Corps of Engineers
- Scanned and rectified USGS Topographic Quads

4. CONCLUSIONS AND RECOMMENDATIONS

At least five major hurdles need to be addressed before GIS crisis mapping becomes widespread during and after disasters. GIS in their present formats are still difficult to operate unless run by technocrats. They need to be made more accessible and easier to operate so they can be implemented for mapping by any intelligent non-specialist. GIS need to be able to integrate data structured in different formats. An inventory of existing GIS needs to be conducted in order to identify which are most appropriate for use in an emergency situation when managers are under stress and time constraints. Emergency personnel need to be trained in what crisis maps might be most helpful to prepare during or after an emergency and what GIS to use. Hardware needs to be developed to build a self-contained GIS mapping capability that would run with the use of generators or batteries and would be able to handle large datasets. To solve many of these problems perhaps a whole new GIS needs to be designed.

In response to Hurricane Andrew, the involvement of private resources highlighted the need for GIS support in emergency mapping. The work of Digital Matrix Service, Inc. clearly demonstrated that by having a GIS-backed mapping capability available, the demand for crisis mapping grew rapidly. The issue of the costs of supplying communities or counties in disaster prone areas with hardware to provide GIS crisis mapping support in an emergency can be put into perspective from this quote from the Wall Street Journal, "the price of a unit of processing power on a chip has fallen by half roughly every 18 months ever since the last 1960's invention of the microprocessors which combined on one piece of silicon groups of circuits that can be programmed." (6)

5. REFERENCES

1. International Decade for Natural Disaster Reduction. (DNDR, 1990-2000). 1987. United Nations, New York.

2. National Research Council, 1987. Confronting Natural Disasters: an International Decade for Natural Hazards Reduction. Washington D.C.: National Academy Press. pp. 1-67.

3. Dymon U.J. and N. Winter, 1992. "Mapping a Technological Disaster."In: Natural and Technological Disasters: Causes, Effects, and Preventive Measures, S.K. Majumdar, G.s. Forbes, E.Q. Miller, and R.F. Schmalz (eds.). The Pennsylvania Academy of Science, pp. 421-439.

4. Bair, F., Elloitt, E. and B. Ruff, 1992. Hurricane Andrew: Storm of the Century, Fort Lauderdale: Dade Book Company, p. 3.

5. Digital Matrix Services, Inc. Daily Log, Reporting Map Use in the Disaster Field Office between September 6, 1992 and October 10, 1992.

6. "Computer Industry Divides Into Camps of Winners and Losers," The Wall Street Journal, February 4, 1993, Vol. LXXIV No. 73, p. 1.

This research was supported by the Natural Hazards Research and Applications Information Center in Boulder, Colorado.

MUNICIPAL WATER SUPPLY DEVELOPMENT FOR LUBBOCK, TEXAS: AN ONGOING QUEST

Otis W. Templer
Department of Geography
Texas Tech University
Lubbock, Texas 79409

Lloyd V. Urban
Water Resource Center
Texas Tech University
Lubbock, Texas 79409

Dan A. Hawkins
Director of Water Utilities
City of Lubbock
Lubbock, Texas 79457

1. INTRODUCTION

Lubbock, with a population of 187,000, is the principal city of the semi-arid southern High Plains of Texas. Here on the South Plains, a part of the Llano Estacado, the extensive, nonrenewable Ogallala aquifer is the major water resource. Surface drainage is very poorly developed and only a few shallow canyons, or draws, cut back into the eastern edge of the Llano Estacado. Most runoff accumulates in the thousands of shallow playa lake basins which dot the Llano Estacado, where most of it soon evaporates. Thus, the South Plains provides little surface flow to the area east of the Caprock Escarpment, and opportunities to develop local surface water sources are quite limited. In early Lubbock, founded in 1891, most homes and businesses had their own windmill wells tapping the Ogallala aquifer. A city water system, established after the turn of the century, relied solely on local wells until the 1950s, when the city began acquiring groundwater rights, now totalling about 82,000 acres, about 60-70 miles to the northwest. By the 1960s, Lubbock had joined ten other High Plains towns and cities in the Canadian River Municipal Water Authority formed to exploit the newly completed Lake Meredith on the Canadian River. A lengthy pipeline aqueduct from this reservoir now provides the bulk of municipal water supply, and the City's extensive groundwater resources are used mostly to seasonally supplement this primary source. In the early 1970s, Lubbock began searching for a second dependable surface water source, eventually culminating in the Justiceburg Project on the South Fork of the Double Mountain Fork of the Brazos River 50 miles to the southeast.

This paper, a study in applied water resource development: 1) summarizes the historical evolution of Lubbock's municipal water supply, including the transition from total reliance on groundwater to heavy dependence on surface water sources, 2) examines Lubbock's newest surface water source, still undergoing development on the Brazos River, and 3) discusses the many physical, economic, and legal-political problems the City of Lubbock has encountered and largely overcome in its ongoing quest to assure an adequate municipal water supply well into the twenty-first century.

2. HISTORIC MUNICIPAL WATER DEVELOPMENT

Surface water first attracted people to the Lubbock area which is located near the juncture of two shallow draws cutting back into the Llano Estacado, the Yellowhouse and North Canyons. In Yellowhouse Canyon, springs draining the Ogallala formed a small lake which existed until the 1920s, and overflow from the lake and other springs kept a small stream of water running down the Canyon. Paleo-Indians occupied the Lubbock Lake Landmark, as it is now called, for 11,000 years or more, as did the later Comanche Indians. The site was also known to Spanish explorers, who marked it as a "place of water" on their maps. In the 1870s-80s this water source was used by the military, buffalo hunters, and the first ranchers and settlers in the region. (7) In 1891, small competing settlements which had developed merged to form Lubbock on the south side of Yellowhouse Canyon. (4)

Away from the canyons on the upland surface, the Ogallala's water table was generally too deep for dug wells, but professional well-drillers were able to tap the aquifer with many shallow windmill wells. Connor notes that "the most striking aspect of early Lubbock was that it appeared to be a forest of windmills, all spinning, clanking and pumping at the same time." (5) Lubbock's focal point, the Courthouse Square, had a windmill well and water trough, which for two decades was the town's nearest approach to a public water system. Nearly every home and most businesses relied on their own private windmill wells and a raised storage tank for water supply.

After the turn of the century, as Lubbock's population approached 2,000, the need for a public water supply system became obvious. The City, incorporated in 1909, constructed the Lubbock Water Works in 1911 and early 1912. Water mains were installed in 1911 and a separate Water Works Department was established in 1912 when a lone operator was hired. A 206-foot deep well pumped with compressed air, a 150,000 gallon ground storage reservoir, and a 75,000 gallon raised steel water tank comprised the new water system, but it was several years before many citizens abandoned their private windmill wells and were connected to the city water system. (2) (6) A second well utilizing a 16-foot windmill for pumping was drilled in 1914; another drilled in 1917 was the first to use a turbine pump driven by an electric motor. However, until 1925, when Lubbock's population reached about 12,000, the City's water was supplied by a single well. In the 1920s, Lubbock's leaders urged that additional local well sites be acquired for future use, and afterward many new wells were drilled. By 1930, the water systems included five wells and 50 miles of water mains. In the 1930s, the City also purchased land encompassing the Lubbock Lake Landmark site for a future surface water reserve, but municipal and agricultural groundwater withdrawals had already lowered the water table below the lake bed and springflow ceased. (7) Thus, this historic site, so important in attracting people to the Lubbock area, never provided any surface water for the City.

Prior to 1955, Lubbock still obtained its water supply from wells in or adjacent to the City. By this time, the City, whose population had grown to 100,000, had acquired and developed approximately 5,000 acres of local groundwater rights and had drilled 92 wells, all within a six-mile radius of the Courthouse. (8) During this time it became apparent that further expansion of a municipal water system dependent on local groundwater was no longer feasible. Irrigated land in the Lubbock area had become too valuable, making the cost of purchasing large adjacent tracts of groundwater rights prohibitive and forcing the City to look beyond the local area for a new water source. (10) (11)

2.1 DISTANT WELL FIELDS

Because of the scarcity of surface water sources in the surrounding area, Lubbock began acquiring additional groundwater rights in more distant areas, in particular the Sandhills of Bailey and Lamb Counties 60 to 70 miles to the northwest (see Figure 1). The Sandhills, encompassing about 200,000 acres, extend for 60 miles in an eastward direction from the New Mexico border. (8) Because of the very sandy soil and dune topography, there is very little runoff, and recharge rates to the Ogallala aquifer are especially high. In the 1950s, the Ogallala aquifer in this area had a high water table because it had been used almost exclusively for ranching and had not been developed for irrigation since the sand dunes and sandy land were unsuitable for furrow irrigation. (10)

In 1952, the City was able to purchase groundwater rights to 53,190 acres of land in the Sandhills in the western part of Bailey County, 70 miles northwest of Lubbock, much more cheaply than the cost of groundwater rights in irrigated areas surrounding the City. A short time later, the City obtained additional groundwater rights in eastern Bailey County and western Lamb County. In 1953, Lubbock bought 2,060 acres of groundwater rights near Shallowater about 14 miles to the northwest, adding to its groundwater reserves. By 1954, wells had been drilled and a pipeline constructed to the Shallowater Well Field.

During 1956-57, the pipeline was extended from the Shallowater Field to a new pumping station in eastern Bailey County. By 1957, the City had purchased more than 76,000 acres of groundwater rights in the Sandhills and Shallowater Well Fields. (8) Subsequent smaller acquisitions have raised the present total to approximately 82,000 acres.

By 1977, 85 wells were in operation, covering less than one-sixth of the City's distant groundwater holdings. Since then, a total of some 270 wells have been placed in service. Initially, it was estimated that the area covered by these water rights contained about 3.3 million acre-feet of water with over 1.5 million acre-feet recoverable. Later estimates reduced the recoverable reserves to slightly over 1 million acre-feet, and most experts agree that the full use of these well fields would exhaust them prior to 2020. (8) However, since development of the Canadian River System, discussed in the next section, water is taken from the well fields only when it is needed to meet the peak demand of the summer months. The City estimates that annual recharge in the Sandhills Well Field is bout 0.5 inches, or approximately 3,400 acre-feet per year. (3) During the past seven years, groundwater production from the Sandhills Well Field has averaged over 5,000 acre-feet per year, well above the recharge rate. In order to conserve the resource and minimize water level declines, the City has chosen, when possible, to limit future groundwater withdrawals to equal the estimated annual recharge. The small Shallowater Well Field has been inactive for the last several years due to declining water quality, but it may be reactivated in the next five years to supplement the Sandhills groundwater supply.

Despite relatively high recharge rates in the Sandhills and though only a fraction of the total reserves have been extracted so far, the Ogallala aquifer there remains a nonrenewable resource. Under Texas law, landowners are able to pump and use percolating groundwater beneath their land despite possibly depriving adjacent landowners of water supply. (12) Thus, this reliance on the "strict common law" rule poses a potential threat to Lubbock's well fields. Though the fields are within the boundaries of High Plains Underground Water Conservation District No. 1, the district does not control production of groundwater, except as accomplished through well-spacing regulations. In the 1950s, when most of these water rights were acquired, it was believed that the Sandhills would always be unsuitable for irrigated agriculture. Since the 1970s, changing technology has drastically altered this situation through the development and widespread use of center-pivot and sprinkler irrigation. The Sandhills Well Field is now flanked on the north and south sides by a significant amount of irrigated agriculture. Further competition for area groundwater stems from Southwestern Public Service Company and the City of Littlefield, both of which own several sections of groundwater rights in the Sandhills east of Lubbock's well field. Though the amount of water pumped for irrigation, industrial use, and public supply in this immediate area has not been determined, it is believed to be several times the natural recharge to the Ogallala aquifer. Further significant expansion of Lubbock's Sandhills water rights is unlikely because of: 1) the conversion of most former ranch land to irrigation, 2) a significant increase in the value of these water rights, and 3) the acquisition of large tracts of groundwater rights by public utilities and other area towns. (9) (10)

2.2 CANADIAN RIVER WATER DEVELOPMENT

Even before acquiring its distant well fields in the Ogallala, Lubbock wanted to develop a surface water source and focused on the Canadian River 160 miles to the north. Unlike the nearer Brazos, Colorado, and Red Rivers, which rise on the semi-arid Llano Estacado, the Canadian River has its headwaters in the mountains of New Mexico and a much greater sustained flow. In 1948, the U.S. Bureau of Reclamation (USBR) conducted a feasibility study for developing the Canadian for municipal and industrial water supply. Lubbock soon joined 11 other High Plains towns and cities in forming the Canadian River Project Organization Committee, which later became the Canadian River Water Users' Association, to promote this development. Also in 1950, New Mexico, Texas, and Oklahoma entered into the Canadian River Compact apportioning the river's waters, and the Canadian River Project was approved as a federal reclamation project. (1)

The Texas Legislature created the Canadian River Municipal Water Authority (CRMWA) in 1953 to develop this project. Eleven of the 12 cities, including Lubbock, ratified membership in the Authority. In 1956, the CRMWA applied for and received a Texas Water Commission (TWC) permit to appropriate 100,000 acre-feet from the Canadian River for municipal purposes and 51,200 acre-feet for industrial use. (14) After early efforts of the CRMWA to fund the project failed, in 1957 the CRMWA requested the USBR to begin action toward construction of the project with federal funding. By 1960, the CRMWA and its member cities had reached agreement on the amount of water each would receive and an apportionment of costs. Local elections in these cities overwhelmingly approved the project. (1) The 11 cities served by the Canadian River Project are Amarillo and Lubbock, the largest cities, plus Pampa, Borger, and Plainview in the north and Levelland, Brownfield, Slaton, Tahoka, O'Donnell, and Lamesa in the south. Lubbock and Amarillo were each allotted 38,170 acre-feet of water annually. (9) In 1962, the USBR began work on Sanford Dam to impound Lake Meredith and completed the dam in 1965 (Figure 1, supra). Construction of an aqueduct system began in 1963, and by 1967 the entire Canadian River Project was completed. (15) The distribution system consists of 322 miles of pipeline, 10 pumping stations, and several regulating reservoirs. Lubbock and the other six towns on the South Plains are all served by a single water treatment plant built just north of the City.

As a reclamation project, Lake Meredith is unique because it is confined exclusively to providing water for municipal and industrial purposes. The purpose of Lake Meredith was to impound relatively low quality surface water to blend with higher quality groundwater to attain a water supply appropriate for municipal use and human consumption. (13) Each city decided to continue to utilize their existing groundwater wells for blending and to augment the supply from CRMWA during the seasonal peak usages, but water from Lake Meredith has historically constituted the largest portion of each member city's water supply. Recently, due to the slowly degrading quality of the groundwater in most areas of the southern High Plains and more stringent limits imposed by the Safe Drinking Water

Act, several of the cities in the south have been forced to depend wholly on the supply from Lake Meredith to meet their water demands.

The Canadian River system, once viewed as a long-term solution to Lubbock's water supply problems, has not been without problems. Specifically, they are: 1) low inflow and reduced storage in Lake Meredith causing the CRMWA to reduce allocation to member cities, 2) continuing disputes between the states over the Canadian River Compact, and 3) most serious, as a result of natural phenomena and human activities in the watershed, water flowing into Lake Meredith is becoming increasingly saline. (10) (11)

Under ideal conditions, Lubbock should receive its full contract allocation amounting to about 12.5 billion gallons per year. However, dry years and the falling lake level caused the allocation to be cut to 90 percent in 1979. It was further reduced to 80 percent in 1982 and 1983. Since then, allocations have fluctuated between 80 and 90 percent and for the last few years have been at 80 percent, forcing Lubbock and some other CRMWA cities to more rapidly delete their nonrenewable groundwater reserves. Fortunately, Lubbock has been able to purchase surplus water from another member city which does not use its full allotment.

Considerable controversy surrounds the Canadian River Compact which limits the amount of water that New Mexico and Texas may keep in storage reservoirs. In 1986, New Mexico raised the spillway level and expanded the storage capacity of Ute Reservoir, a short distance west of the Texas border. Texas and Oklahoma, already concerned about reduced streamflow and upstream reservoir storage, maintained that these modifications gave Ute Reservoir 35,000 acre-feet more storage capacity than is permitted under the Compact. A federal Special Master has recommended that New Mexico must release additional water to Texas and Oklahoma for the next several years. (11)

Salinity of Canadian River water is becoming an increasingly severe problem which must be resolved if maximum beneficial use is to be made of the Lake Meredith water supply. Since the lake began filling in 1965, it has become more saline due to salt cedar infestation, oil field brine, and the high chloride content of geologic formations the river traverses. A USBR study has discovered that an artesian brine aquifer below Ute Reservoir in New Mexico is responsible for discharging about 70 percent of the chlorides and sulfates into the river. Chloride levels as high as 400 mg per liter have been recorded for Lake Meredith water, much higher than the Environmental Protection Agency's recommended level of 250 mg per liter. (10) Such high salinity levels caused the TWC to consider downgrading the water quality rating for most CRMWA member cities, including Lubbock, and the CRMWA to investigate purchasing groundwater rights in the Panhandle to provide water for diluting the salt content. Lubbock was recently successful in convincing the TWC to establish a two-tier rating system which will allow it and other West Texas cities who meet all "primary drinking water standards" to maintain their approved rating and status.

The Lake Meredith salinity problem could be corrected if brine were pumped from the aquifer to lower the water table, and then disposed of through deep injection wells some distance from the river. This would cost an estimated $3.6 to $9 million, depending upon the number of wells needed to correct the problem. Since the mid-1980s, the CRMWA has sought federal funds to combat the salt problem and to authorize the USBR to design and oversee the installation of a Lake Meredith Salinity Control Project. (11) Finally, in late 1992, congressional authorization was achieved, but federal funding still must be secured. Even when the Salinity Control Project is completed and in operation, there may be no water quality improvement for at least a decade because of the heavy buildup of chlorides and sulfates in the river bed. (13)

In 1958, Lubbock's average daily water consumption was 17 mgd, with peak demand consumption of 38 mgd. By 1991, average consumption had doubled to 36 mgd and peak daily usage to just under 80 mgd. Water from Lake Meredith and the well fields will be incapable of providing enough water to meet the current projected annual demand of approximately 43,000 acre-feet by the year 2000 or shortly thereafter.

3. THE JUSTICEBURG PROJECT

Since at least the early 1970s, Lubbock had been searching for a second surface water source, preferably one that would yield at least 40,000 acre-feet of water per year and ideally as much as 60,000 acre-feet. (9) The Wichita River, a tributary of the Red River, as well as streams in Hartley County northwest of Amarillo, were all considered. However, a 1971 preliminary study identified the nearby North Fork of the Double Mountain Fork of the Brazos River, into which Yellowhouse and North Canyons drain, and the South Fork of the Double Mountain Fork, both in Garza County, as the most likely sources. The best potential dam site was several miles below the juncture of the North and South Forks, but thick layers of exposed gypsum there made the site unsuitable. The White River Municipal Water District already held a TWC permit for the Post site on the North Fork, which it was estimated could provide an annual yield of 10,200 acre-feet. The watershed of the Justiceburg site on the South Fork (Figure 1, supra), though encompassing only 394 square miles, was estimated to be capable of yielding 29,900 acre-feet. (13) Together, it was thought these reservoirs could provide as much water as Lubbock's annual Lake Meredith allocation. (16) However, there were many questions about these potential sources, the major reason they had not been developed previously. The TWC was of the opinion the that Llano Estacado contributed little runoff to the east and inconsistent stream gauge readings made it uncertain how much water might be available. Also, the presence of gypsum and other salts in the geologic materials east of the Caprock caused added concern over water quality. In particular, the Post site on the North Fork is known to be high in chlorides and sulfates, and water there might not be of sufficient quality for municipal use. Finally, the time-consuming, difficult, and expensive process of acquiring water rights, a permit for dam construction for the Justiceburg site, land acquisition, and other required environmental permits and licenses had to be overcome.

In 1977, Lubbock voters approved $5.25 million in a bond election to be used to purchase land and right-of-way for the Justiceburg site. Feasibility and engineering studies were conducted during 1978-1979, and it was predicted that the Justiceburg Project could be completed by the mid-1980s. In 1979, the City's application to the TWC for a water rights permit was immediately contested by six area landowners and an energy company holding oil and gas rights around the site. A TWC hearings examiner held six months of hearings in 1982-1983, the longest ever conducted by the TWC, before eventually recommending approval of the permit. In 1984, the TWC granted Lubbock's permit for reservoir construction and the appropriation of 35,000 acre-feet of water per year. Opponents appealed the TWC decision to a district court where the case remained for two more years. Finally, in 1986, the City reached agreement with the opposing landowners and the case was settled. (11) The City agreed to purchase 4,751 acres from the landowners of the almost 6,000 acres needed for the reservoir site. Finalizing these sales required another two years. Though a small oil field crossed the reservoir site, the City did not begin acquiring mineral interests from the landowners until 1989. In the earlier stages of the project the City, its engineers, and consultants never considered the acquisition of the mineral rights in the area to be a major stumbling block. This was primarily due to the age of the oil field, because pumpage from the field had been reduced to marginally economic levels, with most wells plugged and abandoned, and earlier assumptions that drilling could be accomplished from barges or man-made islands in the lake. Recently, it has been determined that economic, environmental, and water quality concerns virtually rule out the possibility of such practices for the mitigation of both present and future working oil wells. The City is now negotiating the purchase of certain outstanding mineral interests, which may add $8 million or more to the cost of the project. Most of the remaining land to be purchased was in the upper reaches of the project site where it would not delay construction of the dam and spillway, and not until 1992 was the last major parcel of land acquired. Thus, following the 1986 court settlement, the City could finally proceed with developing the dam site and securing other required permits, licenses, and approvals.

The application for a Corps-of-Engineers Sect. 404 permit, required for all dams, reservoirs and in cases where other significant changes are to be made in stream flow, was beset by numerous delays and took much longer than originally expected. Coupled with the Sect. 404 permit is the required mitigation for archaeological sites and for displacement of wildlife and destruction of their habitat. Over 70 archaeological sites were identified on the site in an investigation that took several years and cost almost $3 million. In 1990, the City purchased 2,200 acres of land north of the dam site to complete a tract of about 3,000 acres provided for wildlife mitigation.

Though Lubbock had several available options, it chose to contract with the Brazos River Authority (BRA), which already operates several other reservoirs in the Brazos River basin, for construction, maintenance, and operation of the project. Storage in the reservoir will not significantly affect downstream water rights and other lakes managed by the BRA. It is expected to reduce the firm yield of Possum Kingdom Reservoir, the next lake downstream, by only 0.2 percent, and the City agreed to compensate the BRA for this slight reduction. The contract for construction of John T. Montford Dam was awarded in early 1991, site preparation and actual construction were soon underway, and the dam site was dedicated in April of that year. The reservoir, named Lake Alan Henry, will include almost 2,900 surface acres, extend 11 miles upriver, have a storage capacity of almost 116,000 acre-feet, and will be the largest lake in the Lubbock area. The City has set aside approximately 580 acres for parks and public access in the Samuel W. Wahl Recreation Area near the dam site which is already open to public use. Easy public access assures that the lake will see considerable use for water-based recreation. Eventually, the nearby 3,000 acre wildlife mitigation area will be open to the public for controlled hunts and non-consumptive enjoyment. At present, Montford Dam is about 90 percent complete, but closure of the dam and filling of the lake, which might take from two to six years, may be delayed until the dispute over outstanding mineral interests is concluded.

Lake Alan Henry is expected to provide a firm yield of 23.3 mgd, "firm yield" meaning a two-year reserve of water is retained in the lake. However, since this is a supplemental water source, it is not expected that firm yields will be maintained. The City could pump the lake dry if needed, and managed in this way the lake could provide as much as 31.2 mgd. Since the lake will not be maintained at firm yield levels, water-based recreation will obviously be limited during low-water periods. It is expected that the water will be of about the same quality as that from Lake Meredith, however the sources of salinity are so dispersed in the watershed that it will be infeasible to use the same methods of salinity control now being considered at Lake Meredith. Blending Lake Alan Henry water with higher quality groundwater will probably be the most practical and cost-effective method of salinity control. The BRA and the City of Lubbock recently completed a watershed pollution abatement study to determine potential pollution sources. None was found, but this study also assessed water quality from the watershed. Based upon the levels of total dissolved solids, including chlorides, the City has already predicted that the Lake Alan Henry water supply will require blending.

To date, the cost of the reservoir project, including legal fees, Sect. 404 permit requirements, land acquisition, and dam construction has totalled almost $61 million. Water rates have already been increased to cover the debt service on this large capital outlay. Other parts of the project will ultimately involve construction of a 55-mile long pipeline aqueduct, four pumping stations to lift the water about 1,000 feet to Lubbock, and construction of a new water treatment facility with terminal storage reservoir and pump station. Total Justiceburg Project costs are estimated to be about $140 million (in 1990 dollars), not including possible inflated costs if the City decides it must purchase outstanding mineral interests. In any event, it promises to be an expensive project and will necessitate water rate increases by the City after the turn of the century. Though current projections show that Lake Alan Henry's water may not be critically needed until about 2010, it will be at least the turn of the century before water from the Justiceburg Project reaches Lubbock residents, assuming that all current permitting, land acquisition, and mineral rights disputes are resolved and no further unforeseen problems arise.

4. CONCLUSIONS

This paper has discussed how people were first attracted to the site of Lubbock by surface water in Yellowhouse Canyon, but how after the City's founding for over seven decades it depended solely on groundwater from the Ogallala aquifer. Since deliveries from Lake Meredith on the Canadian River began in early 1968, the City has relied primarily on surface water in an attempt to conserve its nonrenewable groundwater reserves. Lubbock will continue its heavy dependence on surface water into the next century, following completion of the Justiceburg Project, and nonrenewable groundwater will be used only for blending purposes and for peak season supplemental water supply.

This paper has shown the many physical, economic, and legal-political problems Lubbock has encountered and largely overcome in developing its municipal water supply. In addition, it clearly indicates that gaining approval for any new reservoir project in water-short West Texas faces the most difficult odds. Despite its location in a semi-arid agricultural region where there is considerable competition with agriculture, industry, and other cities for dwindling groundwater and where little surface water is available, past long-range planning and action has enabled Lubbock to meet its water needs. The City has endured no water rationing caused by inadequate water supply, unlike most other major cities in Texas, whether in more humid or drier regions. When Lake Alan Henry comes on line, the City's combined water sources will be capable of supplying a population of 350,000. Even now, Lubbock continues to look at still other water supply options which it may need if growth and water demand continue as projected, among them desalinization and reclamation and reuse of wastewater. While these options are unattractive and present various economic, legal-political, or water quality problems, they provide further evidence that the City of Lubbock is continuing to explore all possible alternatives to assure an adequate municipal water supply in the years ahead.

5. REFERENCES

1. Brannon, R.G., 1973. An Economic Study of the Canadian River Project from Its Inception through 1971. Austin, TX: Unpublished M.B.A. Thesis, Texas Tech University, 103 p.

2. Brief History of Water Supply System for City of Lubbock, n.d. Lubbock, TX: Water, Reference File in the Southwest Collection, Texas Tech University, n.p.

3. Comprehensive Ground Water Management Study for the City of Lubbock,

1990. Midland, TX: Geraghty and Miller, Inc., v.p.

4. Connor, S.G., 1962. "The Founding of Lubbock." In: A History of Lubbock, L.L. Graves (ed.). Lubbock, TX: West Texas Museum Association, pp. 68-97.

5. Connor, S.V., 1962. "The New Century." In: A History of Lubbock, L.L. Graves (ed.). Lubbock, TX West Texas Museum Association, pp. 98-126.

6. Graves, L.L., 1962. "Health, Medicine and Sanitation." In: A History of Lubbock, Graves (ed.). Lubbock, TX West Texas Museum Association, pp. 173-191.

7. Holden, W.C., 1962. "Indians, Spaniards and Anglos," In: A History of Lubbock, L.L. Graves (ed.). Lubbock, TX: West Texas Museum Association, pp. 17-44.

8. Lubbock and Its Water, 1957. Lubbock, TX: City Engineering Department, 10+ p.

9. Public Works Facilities Study, 1977. Lubbock, TX: Report No. 6 of Lubbock Comprehensive Plan, Department of Planning, 139 p.

10. Templer, O.W., 1990. "Municipal Water Supply for a High Plains City: Lubbock, Texas." Forum of the Association for Arid Lands Studies, 6:59-67.

11. Templer, O.W., 1991. "Legal-Political Issues Affecting the Municipal Water Supply of Lubbock, Texas." Water Management of River Systems, Proceedings of the 27th American Water Resources Association Conference, pp. 393-394.

12. Templer, O.W., 1992. "Texas Groundwater Law: Changing a Venerable Legal Doctrine." Papers and Proceedings of the Applied Geography Conferences 15:1-9.

13. Texas Department of Water Resources, 1977. Continuing Water Resources Planning and Development for Texas. Austin, TX: Texas Dept. of Water Resources, v.p.

14. Texas Water Development Board, 1966. A Summary of the Preliminary Plan for Proposed Water Resources Development in the Canadian River Basin. Austin, TX: Texas Water Development Board, 32+ p.

15. U.S. Department of the Interior, Bureau of Reclamation, 1971. Canadian River Project. Washington, DC: U.S. Bureau of Reclamation, 8 p.

16. "Water: Municipal/Industrial," 1980. Greater Lubbock 53(10):7.

GIS-BASED SHIFT-SHARE ANALYSIS: APPLICATIONS IN METROPOLITAN ECONOMIC MODELING

Daniel Z. Sui
Department of Geography
Texas A&M University
College Station, Texas 77843

1. INTRODUCTION

The 1980s saw an explosive growth of GIS applications in urban planning processes in both developed and developing countries. GIS technology has greatly facilitated information management, information retrieval, and automated mapping of various spatially referenced data in different stages of urban planning. Although the benefits and advantages of using GIS in urban planning are well documented in the literature (10, 13), urban planners have increasingly recognized the limitations of the current generation of GIS technology as a tool for planning purposes. (6) In essence, most, if not all, current GIS functions are cartographically oriented. They usually do not reflect what planners do in the planning process. (3) A general consensus is that GIS will be successful only if appropriate interfaces between GIS and various planning models and techniques are established. (14)

The objective of this paper is to demonstrate how GIS can be integrated with conceptual models and analytical techniques for policy impact analysis. Specifically, by setting this research within the context of the Erickson model on the evolution of suburban economy, an extended shift-share analysis has been integrated with a vector-based GIS to model the spatial impacts of new town development in Hong Kong. Policy implications of the modeling results are also addressed in this paper.

2. THE STUDY AREA AND CONCEPTUAL FRAMEWORK

The study area of this research is Hong Kong, a British Crown colony located on the southeast coast of China. With a population of 5.6 million in 1991 and an area of only 1076 square kilometers, two-thirds of which is hilly, Hong Kong's development has always been constrained by the availability of developable land. The overcrowding in the central urban areas, which have developed around the inner part of Victoria Harbor, has long been a problem in the urban planning of Hong Kong. Since 1960, the development of new towns has been pursued by the Hong Kong government to disperse the population from the central urban areas. Up to the present, eight new towns have been developed. Although the original objective of new town development was to disperse the population from central urban areas to outlying peripheral areas, town planners have strived to achieve balanced social development and self-contained employment in the new towns. While the Hong Kong's new town program has been evaluated by scholars from many different perspectives (11, 2, 9), the spatial economic impacts resulting from new town development have barely been discussed in the literature. In contrast to previous research, this author contends that the spatial impacts of Hong Kong's new town development should be understood as manifestations of spatial restructuring and structural transformation in Hong Kong's metropolitan economy.

The conceptual framework of this study is set within the context of the Erickson model on suburban space economy. The Erickson model (4), as later extended by Hartshorn and Muller (7), articulated the evolving patterns of the suburban metropolitan economy in three stages: (1) *Spillover and Specialization.* During this stage, major employment is predominantly concentrated in the central urban areas, but some sectors began to spill beyond the boundaries of the traditional central city. The spillover of employment tends to specialize highly in the limited blue-collar, labor-intensive sectors. (2) *Dispersal and Diversification.* The employment patterns during this stage are characterized by the extensive spatial dispersal and sectoral diversification, reflecting a wider range of activities and greater balance among their component sectors. (3) *Infilling and Multinucleation.* During this stage, employment patterns reflect continued dispersal, but with substantial infilling of jobs in the inner rings of suburbs close to circumferential freeways. An increasing share of employment tends to concentrate in a small number of suburbs with high rise/high-tech business centers. The entire metropolis tends to become reorganized into a set of independent urban realms, and metropolitan business landscapes are characterized not just by the traditional Central Business District (CBD) but by the development of Suburban Downtowns as well. (7) By tying the Erickson model into this research, this paper aims to address the following two questions through the GIS-based shift-share analysis: (1) In which stage does Hong Kong's new town development fit into the Erickson Model? (2) What are the policy implications for the urban planning of Hong Kong?

3. METHODOLOGY

The methodology developed in this paper is a GIS-based shift-share analysis. While the shift-share analysis offers analytical tools for modeling the impact of new town development, the GIS provides the integrated technical environment for the implementation of the extended shift-share analysis.

3.1 SHIFT-SHARE ANALYSIS AND ITS EXTENSION

Originally developed as a tool to describe the differential responses of regions and states to the national economic change, shift-share analysis can be regarded as a technique that uses proportional measures of change to examine the interdependencies of a sub-regional area with the parent area in terms of the major components of growth. (12) Theoretical assumptions behind shift-share analysis are: small economies change in some definable relationships to a large benchmark or parent economy. Original specification of the shift-share analysis decomposes the growth into three major components --- the national growth effect (N_{ij}), industrial mix effect (S_{ij}), and regional competitive effect (R_{ij}):

$$\Delta E_{ij} = N_{ij} + S_{ij} + R_{ij} \qquad (1)$$

(See APPENDIX I for definitions and explanations.)

One of the serious drawbacks of the original shift-share analysis is its inability to account for the strength of the local market and the effect of local industry mixes. In order to account for the strength of the local market and the effect of local industry mixes, Arcelus (1) extended the traditional shift-share analysis by breaking down the regional shift R_{ij} into two components RN_{ij} (regional growth effect) and RS_{ij} (regional industrial mix effect). Thus the shift-share analysis becomes:

$$\Delta E_{ij} = N_{ij} + S_{ij} + RN_{ij} + RS_{ij} \qquad (2)$$

By incorporating the Esteban-Marquillas's concept of homothetic employment (5), RN_{ij} and RI_{ij} are extended as:

$$RN_{ij} = E^{H}_{ij}(P_{nj}-P_{nN}) + (E_{ij}-E^{H}_{ij})(P_{nj}-P_{nN}) \quad (3)$$

$$RS_{ij} = E^{H}_{ij}[(P_{ij}-P_{nj})-(P_{iN}-P_{nN})] + (E_{ij}-E^{H}_{ij})[(P_{ij}-P_{nj})-(P_{iN}-P_{nN})] \quad (4)$$

(E^{H}_{ij} for Homothetic Employment; see APPENDIX I for definitions of symbols in Equations 1-4.)

By making such extensions of the original shift-share analysis, the specializations and growth differentials in major economic sectors within a metropolitan area can be identified. The meanings and interpretations of equations (3) and (4) are contained in Table 1.

TABLE 1

INTERPRETATIONS OF THE EXTENDED SHIFT-SHARE ANALYSIS

	[(Pij-Pnj)-(PiN-PnN)] > 0	[(Pij-Pnj)-(PiN-PnN)] < 0
(Eij-EHij)>0	High Dependence/Lead Area: higher dependence in region j's growth on sector i and an accelerated growth of sector i in region j than the national average.	High Dependence/Lag Area: higher dependence in region j's growth on sector i and a failure of sector i in region j to keep pace with the national average.
(Eij-EHij)<0	Low Dependence/Lead Area: lower dependence in region j's growth on sector i and an accelerated growth of sector i in region j than the national average.	Low Dependence/Lag Area: lower dependence in region j's growth on sector i and a failure of sector i in region j to keep pace with the national average.

3.2 IMPLEMENTATION

The integration of shift-share analysis with GIS is implemented in the following three steps: (1) *Construction of the Database*. Atlas*GIS is used for creating both the spatial database and the attribute database. (2) *Extended Shift-Share Analysis*. Employment in major industrial sectors is extracted and saved as a separate file. The structure of the new attribute file containing each individual industrial sector is modified to include the items shown in equations (3) and (4). The macro math functions in the Atlas*GIS database are used to execute the extended shift-share analysis. The analytical results are related back to the spatial boundary file through the Geographic Index File (Figure 1). (3) *Mapping the Results*. The thematic mapping module of Atlas*GIS is used for mapping the results and cartographic design.

4. RESULTS AND DISCUSSIONS

The GIS-based shift-share analysis is used to examine the spatial specialization and differential growth of three major employment sectors in Hong Kong between 1966 and 1986. As shown in Figure 2a, the High Dependence/Lead areas in manufacturing employment are clearly clustered in three new towns: Tai Po, Sheung Shui/Fanling, and Tuen Mun. The other new towns, including Sha Tin and Yuen Long, also took the lead in the growth of manufacturing jobs, although these new towns did not specialize in the manufacturing sector. In contrast, manufacturing employment in the central urban areas (i.e., areas surrounding the harbor) is characterized by either the Low Dependence/Lag or High Dependence/Lag areas (such as districts Sham Shui Po, Wong Tai Sin, and Kwun Tong). Such a spatial pattern is a clear reflection on the spillover of manufacturing jobs to some of the new towns and the decline of manufacturing activities in the central urban areas. For employment in the commerce sector, growth and

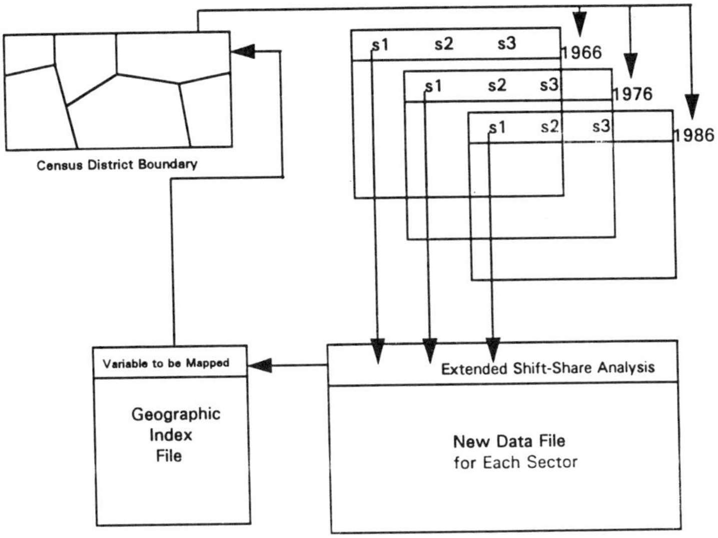

Figure 1. The Implementation of GIS-Based Shift-Share Analysis

concentration in the new town areas are not very significant (Figure 2b), as indicated by the Low Dependence/Lag and a limited area of High Dependence/Lead areas (Yuen Long and Sha Tin). The High Dependence/Lead areas in the commerce sector still agglomerate in the central urban area (Figure 2b). There is no clear dispersal and spillover effect in the commerce sector. Although only two types of areas are identified for the service sector (Figure 2c) -- High Dependence/Lead and Low Dependence/Lag -- it is clear that the central urban areas highly specialized in the service sector took a commanding lead in growth during the study period. In contrast, the new towns exhibited not only a lower dependence on the service sector but also a lag behind the central urban areas in growth.

The development of new towns in Hong Kong has dramatically changed the spatial structure of the city, as evidenced by the emergence of the multi-centered ecological structure. (8) The modeling results in this study, however, indicated that such a multinucleated ecological structure is not accompanied by the multinucleation in the space economy of Hong Kong, as suggested in the Erickson Model. However, the results did confirm that Hong Kong's suburban space economy has reached the Spillover/Specialization stage, and the Dispersal/Diversification stage has not yet been achieved in Hong Kong. Considering the fact that Hong Kong's manufacturing sector has been declining during the past two decades as Hong Kong's economy becomes more and more service oriented (9), the spillover of manufacturing jobs to new towns will not be able to provide enough job opportunities for the residents in the new towns. Many people will have to commute to the central urban areas to seek employment opportunities, particularly for higher paid white-collar jobs. Such a mismatch between the place of residence and the place of work has deteriorated the already congested transportation network in Hong Kong.

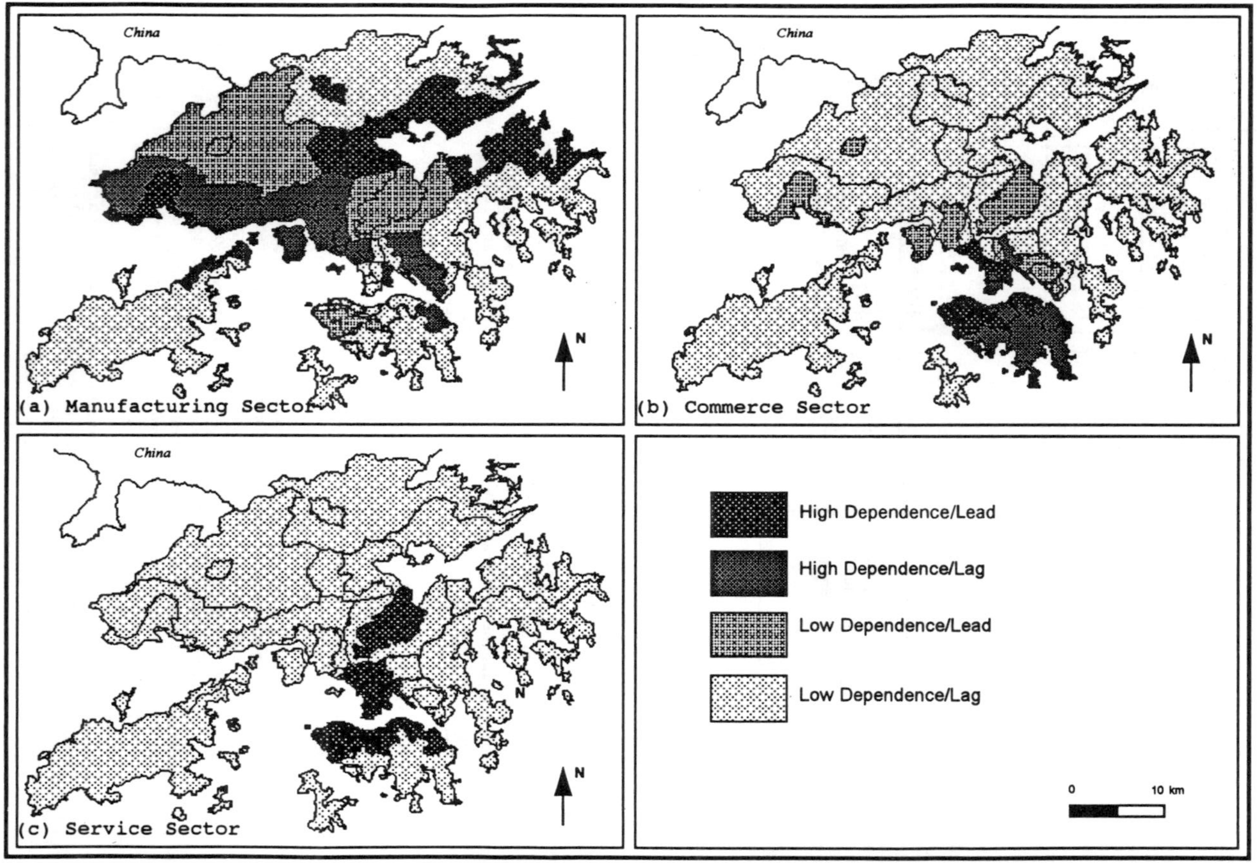

Figure 2. Spatial Specialization and Differential Growth of Major Economic Sectors in Hong Kong, 1966-1986

5. CONCLUSIONS

Geographic Information Systems (GIS) provide geographers and urban planners with a powerful set of tools for manipulation and analysis of spatial information. The functionality of GIS is, however, limited to certain cartographically oriented analysis -- such as overlay and buffer generation. The objective of this paper is to demonstrate how conceptual models (the Erickson model), analytical techniques (the extended shift-share analysis), and technical platforms (GIS) can be integrated for policy impact analysis. The conclusions of this paper can be summarized into two points. First, from a methodological point of view, this paper incorporated an extended shift-share analysis model into a vector-based GIS. This approach takes full advantage of both shift-share analysis and GIS. The extended shift-share analysis empowers GIS users to do more detailed spatial economic development analysis with relatively low cost and less stringent requirements on data while GIS offers the technical platform for the full implementation of the extended shift-share analysis, from database construction and management (both spatial and non-spatial information) to automated mapping and results visualization. Second, from a policy point of view, the public housing-led new town development strategy has not yet brought in the extensive *Dispersal and Diversification* -- the second stage in Erickson's model -- to Hong Kong's space economy. In the absence of concomitant policies to disperse employment, new town development in Hong Kong has resulted in the mismatch between the place of residence and the place of work, and stimulated the polarization between white-collar jobs concentrated in the old urban areas and blue-collar jobs in the new towns. This is an important lesson for other developing countries that are pursuing new towns as a solution to overcrowding in primate cities.

6. ACKNOWLEDGEMENT

My sincere thanks are due to Drs. C.P. Lo, James O. Wheeler, and Jean-Claude Thill for their constructive comments on an earlier draft of this paper. Editorial help from Elizabeth J. Leppman is also grateful acknowledged.

7. REFERENCES

1. Arcelus, F.J., 1984. "An Extension of Shift-Share Analysis. "Growth and Change 15(1): 3-8.

2. Bristow, R., 1989. Hong Kong's New Towns, Hong Kong: Hong Kong University Press.

3. Couclelis, H., 1991. "Requirements for Planning-Relevant GIS: A Spatial Perspective." Papers in Regional Science 70(1): 9-19.

4. Erickson, R.A., 1983. "The Evolution of the Suburban Space Economy." Urban Geography 4(2): 95-121.

5. Esteban-Marquillas, J.M., 1972. "A Reinterpretation of Shift-Share Analysis," Regional Science and Urban Economics 2(3): 249-55.

6. Harris, B., 1989. "Beyond Geographic Information Systems: Computers and the Planning Professional," Journal of American Planning Association 55(1): 85-90.

7. Hartshorn, T.A. and P.O. Muller, 1988. "Suburban Downtowns and The Transformation of Metropolitan Atlanta's Business Landscape," Urban Geography 9(4): 373-95.

8. Lo, C.P., 1986. "The Evolution of the ecological Structure of Hong Kong: Implications for Planning and Future Development," Urban Geography 7(4): 311-35.

9. Lo, C.P., 1992. Hong Kong, London: Belhaven Press.

10. Martin, D., 1991. Geographic Information Systems and Their Socioeconomic Applications, London:Routledge.

11. Phillips, D.R. and A.G.O. Yeh (eds.), 1987. New Towns in East and South-east Asia, Oxford, England: Oxford University Press.

12. Stevens, B.H. and C.L. Moore, 1980. "A Critical Review of the literature on Shift-Share as a Forecasting Technique," Journal of Regional Science 20(4): 419-38.

13. Worrall, L. (ed.), 1992. Spatial Analysis and Spatial Policy Using Geographic Information Systems, London: Belhaven Press.

14. Yeh, A.G.O. and M. Batty, 1990. "Applications of Geographic Information Systems in Urban and Regional Planning (Guest Editorial)," Environment and Planning B 17(1): 1-7.

APPENDIX I. DEFINITIONS OF SYMBOLS IN EQUATIONS 1-4

1. i represents an individual industrial sector.
2. j represents a region.
3. n represents all industrial sector.
4. N represents the nation (or the benchmark economy in shift-share analysis.
5. E_{ij} is the employment in sector i in region j.
6. P_{nN} is the percentage change of total employment in all the industrial sectors in the nation.
7. P_{iN} is the percentage change of total employment in sector i in the nation.
8. P_{nj} is the percentage change of total employment in all the industrial sectors in region j.
9. P_{ij} is percentage change of total employment in industrial sector i in region j.
10. N_{ij} is the national growth effect, the part of ΔE_{ij} attributable to the growth of the national effect. $N_{ij} = E_{ij}*P_{nN}$
11. S_{ij} is the industry mix effect, the part of ΔE_{ij} attributable to differences between the industrial composition of region j and that of the nation. $S_{ij} = E_{ij}*(P_{iN}-P_{nN})$
12. R_{ij} is the regional competitive effect, the part of ΔE_{ij} attributable to the differences in the growth rate of sector i at regional and national levels. $R_{ij} = E_{ij}*(P_{ij}-P_{nN})$
13. ΔE_{ij} is the change of employment sector i in region j within a period of time. In this paper, the temporal span is 1966-1986.
$\Delta E_{ij} = N_{ij} + S_{ij} + R_{ij}$
14. E^H_{ij} is the homothetic employment in sector i in region j, which is defined as the employment that sector i of region j would have if the structure of the employment in such region were equal to the national structure. $E^H_{ij} = E_{nj} * (E_{iN}/E_{nN})$, where E_{nj} is the total employment in all the industrial sectors in region j. E_{iN} is the total employment in sector i in the nation. E_{nN} is the total employment in all the industrial sectors in the nation.
15. RN_{ij} is the regional growth effect, the part of ΔE_{ij} which is attributable to the growth of region j. $RN_{ij} = E^H_{ij}(P_{nj}-P_{nN}) + (E_{ij}-E^H_{ij})(P_{nj}-P_{nN})$
16. RS_{ij} is the regional industrial mix effect, the part of ΔE_{ij} that is attributable to industrial composition.
$RS_{ij} = E^H_{ij}[(P_{ij}-P_{nj})-(P_{iN}-P_{nN})] + (E_{ij}-E^H_{ij})[(P_{ij}-P_{nj})-(P_{iN}-P_{nN})]$

LAND USE PLANNING AND REGULATIONS IN BRITAIN & CANADA - SOME ASSOCIATIONS WITH LOCAL DEMOCRACY

Alan Hallsworth
SIRC University of Portsmouth
PORTSMOUTH PO1 3HE
UK

1. INTRODUCTION

The influence of land use planning and regulations in determining retail structure is a vast topic to cover for any one country, let alone two. The main part of this paper, therefore, concentrates on the effects of such legislation on just one retail form - the Regional Shopping Centre (RSC). Some attempts are made to contrast Britain and Canada and the paper then looks forward to the prospects for change under new planning guidelines for the UK. The first part of the paper is a summary and modification of earlier work (8) which sought to illuminate some of the issues surrounding the remarkable failure rate of planning applications for large (over 500,000 sq. ft. gross) out-of-town Regional Shopping Centres (RSCs) in the UK, especially during 1989 and to draw comparisons with developments in Canada, where there has already been a massive expansion of such centres (14). The paper is "re-visited" in the sense that the UK planning system has now undergone some revision and it is necessary to re-formulate certain arguments. It was necessary in the earlier paper to abandon a total Canada - UK contrast in favour of comparing one Province with part of the UK. It was decided to choose a Province where comparisons with the UK are maximised and Ontario was selected with this in mind. Ontario utilises planning statutes broadly similar to those in the UK; especially since the 1983 Planning Act. In the context of retailing, particular emphasis should be placed on the role of the Ontario Municipal Board (OMB) which has an 'appeal' function much like that of the Department of the Environment (DoE) in England.

2. PLANNING IN THE UK FROM 1947 TO THE IRON LADY

The 1991 paper argued that during the post-war period of 'Butskellite' consensus politics the UK planning system formulated in 1947 exhibited a certain community of purpose at all levels. A refusal of planning permission at a local level was often supported by the ultimate arbiter, the Minister for Housing and Local Government or, latterly, the Secretary of State at the Department of the Environment (DoE). However, whilst the aims and objectives of planners outside Whitehall may have remained broadly untouched by national politics and market forces, the position within the corridors of power changed dramatically with the election of the Thatcher government in 1979. Nicholas Ridley (Secretary of State at the DoE until July 1989) once declared market forces to be beyond his power; he could not he claimed, control them ... yet most post-war planners would doubtless feel that the control of market forces was one of their *raisons d'etre!*. Fears of a market-led free for all could only have been heightened by the creation, in 1981, of enterprise zones where planning constraints were largely removed and retailers were soon attracted in, even creating RSCs as at Gateshead and Dudley. Hard upon the heels of enterprise zones came proposals for out-of-town RSCs; some of them blatantly seeking Green Belt locations. However, one impediment to a totally *laissez-faire* approach enforced from the top downwards is the reality of UK politics. The Thatcher governments drew their political support dominantly from the affluent southern counties. These are the very areas attractive to proposers of RSCs but where 'not in my back yard' (NIMBY) attitudes are also strong. Accordingly, political tokenism demanded that RSCs should be resisted in the Green Belt areas treasured by the most vocal Conservative voters. Beyond that, the question remained as to what other grounds may be forwarded for refusal of RSC schemes. Because of the interventionist character of the planning appeal system, eighties outcomes tended towards the anarchy of being both place and site specific. This tended to confound the efforts of analysts seeking a pattern in the decisions generated by the planning appeal system. In the earlier paper, this unravelling of UK policy was assisted by examination of policy shifts in Ontario where a clear and established trend of restraint against RSCs had been evident for some time. What, then, were my earlier case studies - and what the conclusions?

3. CANADIAN CASE-STUDY #1: CAMBRIDGE, ONTARIO, 1989

The most recent of the case-studies chosen concerned the City of Cambridge, Ontario (formally GALT) where the City Council had been attempting for years to revitalise the downtown area through the addition of a major department store. However, as soon as one RSC scheme was devised, another, rival, scheme immediately appeared in the form of an extension to an existing mall. At this point the Ontario Municipal Board (OMB) was called upon to adjudicate on the relative merits of the competing schemes. The choice being between the John Galt Centre expanded by 100,000 sq. ft. to incorporate Sears, or the all new Cadillac Fairview scheme a short distance away near the 401 highway. This latter scheme proposed a 500,000 sq. ft. RSC, on a 117-acre site and anchored by an Eaton department store, with foodcourt, hotel, ice-skating rink and other leisure elements. The developers agreed to pay all costs of necessary feeder roads from the 401 Highway.

On 29 December 1988, with a final amendment on 10 February 1989, the OMB supported the city in its decision on the re-zoning and permitted the Cadillac Fairview scheme, *but only the Cadillac Fairview scheme*, to proceed. Consider some of the statements from the OMB decision; 'The Board (OMB) must look at the planning aspects in order to compare the two proposed centres' (16), and 'We believe that this Board should have some concern about the impact on the neighbouring municipalities' (16). This seemed in tune with policies from the early 1970s when Ontario was active in assisting downtowns to fight back against the malls - initiating schemes upon which the city of Cambridge might come to draw in the face of its own new out-of-town RSC. How, then, did this "one mall only" decision contrast with OMB attitudes of 20 years previously?

4. CANADIAN CASE-STUDY #2: ETOBICOKE STORE WARS 1963-68

This case-study, from the 1960s, outlined the circumstances of what came to be known as the Etobicoke Store wars. Again the scenario concerned possible RSC development orientated towards a major highway intersection. This time the location was potentially much more lucrative: being in the Etobicoke area of Metropolitan Toronto. The intersection in question was the junction of Highway 27 on the Queen Elizabeth Way (QEW) west of Toronto. From July 1963 onwards, attempts had been made to develop the 107-acre Sherway site for an RSC ultimately to be in excess of 1 million sq. ft. Gross Leasable Area (GLA). In a 1968 decision on the case, the OMB stated: 'The same arguments which relate to the inability of the Board to interfere with the commercial interests of municipalities, applies with even greater force to the objection of private commercial interests'. In other words, the OMB, in the 1960s, was washing its hands of the issue of commercial competition. The only grounds that the OMB could see for raising objections related to the current state of the road system in the area. The OMB had earlier stated:

> "There is no reason why the citizens of this Province, the expenditure of whose taxes on the QEW and Highway 27 have made this proposed site so attractive, should be forced to accept a level of service below [Grade C traffic flow conditions] on the QEW in order to expedite the building of an RSC".

This, as we shall later see, was a prescient comment!

The Ontario model carries with it a serious message. Ontario did begin to suffer from a proliferation of RSCs and we have seen how these were 'reined-in' by modified decisions of the OMB. Basically, the OMB at last discovered the concept of *too much* shopping in a country already seen as massively over-shopped when compared with the UK (though by 1992 Britain, too was seen as "overshopped"). Basically, as with many other countries, a simple model was followed, thus:

> laissez-faire development >>> oversupply >>> retrenchment.

Now, if *market forces* had been embraced as wholeheartedly in the UK as Enterprise Zone policy implied, would the model be followed? One long-awaited guideline on policy was the parliamentary written answer of 5 July 1985 (12). In this it was suggested that ..."it is not the function of the planning system to inhibit competition among retailers ... not to preserve existing commercial interests as such". Only "cumulative" effects of retail competition on whole town centres would be deemed unacceptable and this only in exception circumstances. The problem is that, as Ontario demonstrated, it is not easy to predict how much competition it takes in order to threaten a whole town centre. Certainly, Ontarians did not want their town centres threatened - but it happened anyway. One could identify sufficient leeway in the *Hansard* statement to encourage developers to put forward schemes with some real expectation of success. Accordingly, proposals for RSCs bounded ahead in 1986, leading Schiller (17) to describe memorably the coming of the 'third wave' of out-of-town retailing.

5. PUBLIC POLICY STATEMENTS

Two subsequent policy decisions by the DoE are worthy of note, for these give some clues as to the true fate of Schiller's 'third wave' despite that liberalising rhetoric. One was the decision of 15 December 1986 to call in proposals in excess of 250,000 sq. ft. Of more significance was the fact that planner began to re-think *their* strategies: Angell (1) stated: ..."Planning authorities must re-appraise their policies which for the most part are directed by the aims in Structure Plans. They must do so in the hope that, if these plans are thoroughly researched and 'hold water' they will be supported by the Secretary of state and his Inspectors on appeal". For the developers the second policy statement was far more serious: Guidance Note PPG6 of 20 January 1988 (revised November 1992).(3) The guidance note stated that planners should take account of forecasts of retail expenditure - which is itself a minefield - but not attempt to set rigid floorspace limits. Most important though, was this:

> Proposals for such developments, which may be well over 100,000 sq. ft. and up to 1 million sq. ft. or more (10,000 - 100,000 sq. m) have no place in the Green Belt ... *Nor are developments on that scale generally acceptable in the open countryside* [my emphasis].

(Incidentally, the 1992 revision wrote of impact on **any** nearby centre). The short spell between the *Hansard* statement and the PPG6 Guideline saw one notable RSC permission - at Thurrock, to the east of London. Overall, however, the scale of refusals was such that Lee (15) was able to speculate that the third wave had broken.

6. THE HAMPSHIRE CASE STUDIES

In the earlier paper, two case studies were taken to show these new planning realities in action - for the proposals were both refused. One was Campus 9, at Whiteley, near Fareham - where it was suggested that the proposal might jeopardise a known expansion scheme for nearby Fareham town centre and "the Whiteley local plan ... only recently adopted and the first alteration to the South hampshire Structure Plan ... recently approved. WITH ... considerable weight to be accorded to these plans".(11) The other major scheme to fail in Hampshire was the 850,000 sq. ft. GLA Adanac park proposal, west of Southampton. This seemed to be an especially interesting test case, given the buoyancy of the local economy. The conclusions were that 'there is ... no justification for making an exception to S1 and S2 (recent County council shopping policies) ... in relation to Eastleigh ... the shopping element of Adanac would cause further harm to interests of acknowledged importance'.(4) One might reasonably conclude that the concept of cumulative impact was being reversed and instead of covering the impact of several schemes on one centre, was taken as the impact of one scheme on several centres.

7. THE CANADA-UK CONTRAST BY 1989

The two Canadian case-studies gave a stable picture through time. There was very much a development-orientated philosophy in the 1960s, perhaps leading to the development of 'killer malls' [Jones and Simmons 1987]. This was followed by a position in the 1970s where Federal and Provincial aid was needed for some downtowns. The RSC development scene slowed and the pattern was more of 'infilling' in smaller centres and return to downtown. In the late 1980s the OMB expressed concern at oversupply, but was happy to permit RSCs in towns as small as Cambridge. In contrast, after decades of post-war restraint, the UK seemed faced with market liberalisation in the 1980s and enterprise zones opened some doors to opportunistic developers. A haphazard period ensued, described by Cole (2) as *ad hocery*. Further proposals quickly gathered pace, with many schemes reaching fruition at the same time. Instead of moving into phase two of the Ontario model (many new permissions given - possibly reaching over-supply) there was a sudden and dramatic *volte-face* with many schemes, including those studied in Hampshire, being refused. The distinguishing feature of the UK experience was that the halt to the tidal wave of proposals was far more sudden and dramatic than anything experienced in North America.

8. THE LESSONS FROM HAMPSHIRE - TO THE REST OF THE UK?

The message from South Hampshire in general and Adanac Park in particular was that RSCs suddenly came to face more development difficulties than might have been supposed at the time of the statement in *Hansard*. The common theme, clearly covered in the then-extant PPG6, is the preservation of open space. Developers, it would seem, need to pay particularly close attention to this. Also, Hampshire moved especially swiftly in the manner outlined by Angell (1). They put in place modern plans that required approval from the Secretary of State. Once those plans were approved, it was difficult to explain why they should so soon be overturned in favour of RSCs. This looks to have been the role-model for future policy nationwide - but drawing causal links requires the passage of time - and more evidence. Secondly, Hampshire is the political heartland of the ruling party. The Conservatives have generally been sensitive to 'home counties' public opinion - currently strongly against commercial intrusions into the countryside.

9. WHAT HAPPENED NEXT?

One consequence of the 1988 budget was that the British Government unleashed a spending spree that caused consumers to "over-reach" themselves (particularly in their property purchases). Caught in this speculative tidal wave, of the few major mall-type developments to be built, two at least saw their owners bankrupted. We may conclude that other UK developers probably had reason to celebrate that the Shopping Centre "boom" did not continue; the losses could have been far worse. Accordingly, advice on how best to approach RSC development in Britain will have to wait for developer enthusiasm to return. Perhaps chastened by the Hampshire experience of having to run through uncomfortable arguments **TWICE**, Government Planning policy itself changed with the 1990 (amended 1991) Town & Country Planning Act.

It also may be that the Government began to see value in distancing itself from the planning process - just as it had taken care to "privatise" - and hence cast adrift - much of Higher Education and the Health Service. No-one can be sure why, but, under the 1990 Planning Act, Government seemed to want to absolve itself from a direct link with future " Great Planning Disasters". The new Act called for tougher formal plans from which variations would be more difficult. So, in theory, appeal to the Secretary of State would be less fruitful. However, planners were also charged with taking account of the arguments of developers. Likewise, the Secretary of State retained his powers to intervene at plan-approval stage if he felt the plan too restrictive.

What is clear, though, is that RSC developers are not in the forefront of the lobbying to be included in new, binding, local plans that are an odd reversion to the inflexibility of 1947 legislation. So, perhaps I am right in speculating that Hampshire set a precedent for the Government to follow. The question remains, of course, as to where the new Planning battles will be fought. The answer, dominantly, is that they lie in the realms of food superstores (mega-stores or hypermarkets).

10. PLANNING ISSUES; WHO GAINS , WHO LOSES? WHO PAYS?

There is, in British Social History, an unbroken link with the status and value of land. In such terms as Landed Gentry we encapsulate a particular notion of status and the history of land is discussed by Ingham (13) and by Hallsworth (9). This paper now discusses one particular set of circumstances that surrounds the development of land for commercial purposes. As detailed in Guy (6) and Hallsworth (7) there are profound consequences that flow from the granting by planning authorities of planning permission for retail development. The site in question immediately leaps in value, bringing windfall profits (planning gain) to the owner and, potentially, the developer. The ramifications of planning gain are well documented in the planning literature and will not be considered further - except in the particular instance of retailing.

11. RECENT RETAIL PLANNING ISSUES

The profound collapse of the UK economy and the particular distress of the construction sector needs little introduction in North America - since it has brought low the major development company Olympia and York, based in Toronto (10). Recession has also thrown into sharp focus the fact that the food retailers are practically the only "growth" sector of the UK economy at time of writing (early 1993). There seem to be two dimensions to this that are currently attracting attention. One is the destruction of open space in urban areas where and when permission for development can be obtained. With the slowdown of the economy, more and more residents claim that it is unnecessary to build more stores. This, coupled with the eternal NIMBY factor, means that such proposals are facing increasing pressure. An adjunct dimension to this is the continuing rapid demise of local democracy in Britain. The period since 1979 has seen a massive centralisation of power as successive Conservative National governments have drawn more and more control unto themselves. This process now seems set to engulf the County Councils. The "Independent on Sunday" of November 22 1992 described - in headline terms - the new system of local government funding as " Local authorities' **last chance** to hold onto their already limited powers to raise their own funds". The article stated " Ministers ... confronted with hostility from the Treasury, which favours centralised control of revenue-raising... Sir Rhodes Boyson, Conservative MP for Brent North... said ' We cannot afford to bring in a measure that will disadvantage people in the South. The test is whether people in my constituency are, on average, no worse off'". Spending levels therefore, as ever, were set by Central Government. The choice or discretion open to Local Government was not one of how much tax to raise locally but of how to stay within a budget set, politically, by their true masters in London.

Where this affects local Democracy (and, as we shall see, "planning gain") is not difficult to discern. Deprived by Central Government of the right to spend (for example they cannot spend the receipts from Council House sales) the Local authorities are strapped for cash. This situation forces local authorities into a sell or starve situation. They must sell council-owned land in order to fund infrastructural development that, only a few years ago, they might reasonably have funded from revenues. There are, often, irreconcilable tensions to be found. An authority may have a firm and clearly-thought-out policy to resist out-of-town development but be strapped for cash for socially-desirable schemes. Largely emasculated by Central Government in terms of their financial independence, they are further threatened by penalties that may force them to financially recompense a thwarted potential developer. So, we can see that Central Government (in ideological terms the "developers friend") need have no doubt that a prudent Council **MUST** allocate some development prospects - or face cash-flow problems. It therefore becomes "safe" to take a hands-off approach.

With the food superstores apparently still pushing for growth, it is their interface with the ragged remnants of Local Democracy that mostly makes the headlines. Often, a local authority will barely receive in "planning gain" sufficient revenue to build adequate access to a new store - access that, without the store, would have been unnecessary. What squares this circle is that some believe we are close to saturation in the food retail market. Increasingly, critics suggest that the growth of food retail stores is unecessary and unsustainable or that there is evidence that the foodstore development process is sustained only by financial illusion (18, 19). It is therefore possible that the concreting-over of scarce urban space is continuing for no good purpose.

12. SQUARING THE CIRCLE - CANADA (ONTARIO) AGAIN

Interest in squaring the circle requires that the attempt be made to bring Canadian lessons into Planning Gain also. I have already instanced situations where the Province of Ontario has enacted legislation that has impacted upon shopping provision. There is also a question of coping with growth since Toronto grew from 1.6m in 1966 to 3.7m in 1986 and the urban area has, therefore, outgrown the limits of Metro. Toronto: refuelling old debates about loss of vital farmland. How should new growth be regulated in the 1990's? Ontario, now run by the NDP, has changed from the days of 30 years of Tory rule. Two factors seem to be growing: one is the "accountability" lobby, the other the "environment" lobby. These throw us back into our headline questions..."Who gains?, who loses?, who pays?".

Welfare economists and others seek to examine "who benefits" when the State pays for infrastructure. One obvious answer is the new developer - as the OMB hinted in its 1965 comments on Etobicoke store wars (above). 1990 brought legislation to tax large commercial developments via CCT (Commercial Concentration Tax, January 1990.) This legislation seemed to link both the accountability and environment lobbies. The background to this policy has to be seen in the light of a 1989 (Liberal) Provincial budget for Ontario that included $2bn in transport infrastructure. The Greater Toronto Area (GTA) was specifically identified as a net beneficiary of Province-wide funding for, in particular, roads. The budget statement therefore included the following:

> "Therefore, beginning in January 1990, owners of large commercial structures and associated parking with gross areas in excess of 200,000 sq.ft. will be required to pay a (CCT) levy of $1.00 per square foot per year". (5)

This statement led inexorably to Bill 46 to establish CCT and which received Royal Assent on December 19th 1989. This was set to yield over $100 million annually with the City of Toronto providing the bulk: much of the residue contributed by Metropolitan Toronto as a whole. The message was totally clear: this bill only applied to the Toronto area because the Toronto area would soak up the bulk of the new transport investment. Especially when the investment took the form of roads then the key net beneficiaries would be outlying shopping malls. The budget statements also noted:

> "Commercial property owners in the (GTA)... realise considerable economic benefit from Provincial expenditures on infrastructure"...and "those who benefit significantly and directly from the provision of new public infrastructure will be required to make a greater contribution to the cost of that infrastructure".

13. PLANNING GAIN - AGAIN

The links to planning gain are all too clear. What is interesting here is that the action was entirely retrospective. The malls over 200,000 sq.ft. were built - the roads were not yet built. Instead of "ad hoc" local access deals, the developers are being faced with a long-term continuing commitment to fund roads. The argument was based on evidence that 67% of all shopping visits to Metro Toronto are still made by car. This is not now regarded as environmentally acceptable in a Province that wishes to see more use of public transport, or at least more high-occupancy car routes and the like. Some roads will be built - but only to serve consolidated built-up areas. CCT, then, is the "stick" elements of a continuing "carrot and stick" policy to lure people onto public transport. It is too early to tell what the true effects will be and whether or not there will be many "unintended consequences" of the legislation. The view from England at this stage is unclear! Certainly, CCT is a complex issue and some argued from the start that it was hastily put together and would be complex to administer. More importantly for present purposes is that what it does show is that public authorities with devolved power are prepared to think about, "who benefits" and "who pays". The Province of Ontario in general (and Toronto in particular) is a fair bet to follow US EPA-type clean-air legislation that has already impacted on the prospect of new malls south of the border. Let us not forget, though, that Ontario is pioneering here. There are still many examples - including ones from other provinces of Canada - where local authorities find themselves with no planning gain and hefty infrastructural bills. It seems that a political decision must be made as to whether or not outlying commercial developments are "a good thing". From that point, the question of "who pays" may be addressed. Provided the authority concerned has the power to raise taxes, the Ontario model may be one of which we will see more in the future. Such initiatives will, however, need to be treated with caution and on their individual merits rather than simply being adopted as proven good practice. It is of some interest that attempts to draw Government attention to EPA-type legislation is being made by UK retailers who favour downtown locations and might wish for protection from out-of-town schemes!

14. W(H)ITHER LOCAL DEMOCRACY IN BRITAIN?

However we view CCT we cannot elude the fact that yet again Canada provides a vivid contrast with Britain. The whole concept of the "local state" is hard to define in a country so apallingly centralised. The iron grip of local control is these days policed by systematic and legally-enforceable cash-starvation. It is legally-enforceable in that Central Government can give itself any powers it likes to enforce any policies it likes in respect of local democracy. It can enforce its centralised views in areas where it holds little or no popular, grass-roots support at any level of the electoral process. The latest twist is that the cash-starvation is now such that even matters of controlling lucrative planning permissions can safely be devolved.

REFERENCES

(1) Angell, R. (1986) The Local Authority View, Proceedings, CBI/Edward Erdman conference: the future of shopping; in or out of town, London: CBI.

(2) Cole, H. (1986) Retail Planning Permissions Proceedings, Summer Annual Meeting, Brighton: PTRC.

(3) Department of the Environment (1988/1992) Planning Guidance Note PPG6 London: HMSO.

(4) Department of the Environment (1989) Decision Letters.

(5) Government of Ontario (1989) "Budget Statement of the Hon. R. F. Nixon" May 17 1989.

(6) Guy C. (1980) Retail location and retail planning in Britain Gower, Farnborough.

(7) Hallsworth A. G. (1988) The human impact of hypermarkets & superstores Gower (Avebury) Aldershot.

(8) Hallsworth A. G. (1991) "Regional Shopping Centres: case studies on recent policy decisions in Canada and the UK" The Service Industries Journal 11.2. pp219-232.

(9) Hallsworth A. G. (1992) The new Geography of consumer spending Pinter publishers, London (Wiley/Halstead).

(10) Hallsworth A G and Bobe MJ (1993) "How the interest rate cat ate the Docklands canary" AREA vol 25 , 1 pp 64-69.

(11) Hampshire County Council (1986) Draft Shopping Policy Document.

(12) Hansard (1985) Retail Developments Parliamentary written answer, 5 July.

(13) Ingham G (1984) Capitalism divided? The City and Industry in British Social Development Macmillan, London.

(14) Jones, K. J. and J. Simmons (1987) Location, Location, Location Toronto: Methuen.

(15) Lee, M. (1989) The Third Wave Breaks Development Economics Bulletin No.4, London: Lee Donaldson Associates.

(16) Ontario Municipal Board (OMB) 1963-1989 Decision Letters.

(17) Schiller, R. K. (1986) Retail Decentralisation: The Coming of the third Wave.The Planner 72, 7.

(18) Shiret T. (1991) "Tesco Plc: a company capitalising too much interest?" Credit Lyonnais Laing 22 April Appold St London.

(19) Wrigley N (1992) " Sunk capital, the property crisis, and the restructuring of British food retailing" Environment and Planning A .

MANAGING LOCATION

Ian Clarke
Department of Retailing and Marketing
Manchester Metropolitan University, UK

1. INTRODUCTION

This paper advocates the need to explore location from the perspective of the retail organisation. The reason for this is that the emphasis of most published work on retail location within geography has been on techniques of locational analysis and consumer store choice. The edited works of both Davies and Rogers and Wrigley illustrate this feature well. (1) (2) Other prominent geographers, such as Brown, have sought to integrate broader theories from other disciplines in search of a more meaningful explanation of institutional change in retailing. (3) This type of work has provided a kernel of theory that geography has lacked, but in order to provide a more powerful framework for understanding and explaining changes in the retail system, an investigation of how locations are managed by retail organisations is sorely needed.

To provide a framework for understanding location in the context of corporate strategy formation, some very useful geographical work has been undertaken, utilizing simple concepts from the marketing literature. (4) (5) Despite the stress laid on location within the retail 'mix' (product, price, promotion and place), marketing texts themselves actually say very little which is of practical use - either to the business manager interested in making the most of a store network, or to the geographer intent on explaining changes in it. Geographers could therefore make a valuable contribution to developing a deeper understanding of the role of place or location within retail strategy. This paper begins to address this issue.

2. THE INFLUENCE OF LOCATION

One of the interesting features in the evolution of geographic work on retailing has been a shift away from the traditional view of the 1970s, which highlighted the growing influence of retailers on the form and operation of marketing channels, as exemplified by Dawson's summary work. (6) This approach compares with the current view of some authors who state rather more explicitly that it is the location of the outlets which is central to successful strategic planning by, and performance of, retail organizations:

> "In the development of competitive strategies, prices can be matched, services can be extended and improved, merchandise may be duplicated, and promotion can be imitated, but a retailer's locational advantages are difficult to assail or neutralize." (7)

The need to define more precisely what location does, is central to evaluating how it can be used. At its most basic, location represents a spatial point of access to a market which itself has a spatially-defined catchment; it is also a potential point of exchange between the customer and retail organization - 'potential' because the location competes with the outlets of other retailers for trade.

Some of the fullest investigations of the role of location in corporate strategy, have been provided by Risto Laulajainen. To begin with the early stages of an organization developing, where the network is relatively small, the spatial expansion through the creation of new outlets introduces uncertainties and potential risks when the organisation is attempting to enter unfamiliar markets. Under these conditions, Laulajainen's work demonstrated the natural hesitance at the early phase of growth can be handled simply by gradual contagious expansion, outwards from the locations and catchments already served. (8) The purpose of managing expansion through location in this way is to retain and control the familiarity of the trading environment, staying close to the customer the organization knows best - both literally and geographically. However, contagious development also has simple financial and logistic rationales, in particular, it enables economies of scale to be gained from the spatial organization of the retail network, through advertising, distribution and management organization, as illustrated by research into the expansion of the UK discount food retailer Kwik Save. (9) Laulajainen and Gadde's study of the expansion of three Swedish retail chains demonstrates further, that in the phases of early corporate retail expansion, locational siting of new stores can be used to avoid competitors. (10) This process of spatial avoidance of competitors has been illustrated by other work on the expansion of bank branches in the USA. (11)

The fine line between adopting a location strategy of avoidance based on free competition and a policy of evasion based on either active or tacit collusion with competing organisations, is a matter raised by Brown. (12) He describes a number of inter-war situations where spatial agreements were common, especially in the grocery and newsagent sectors. Another prime example of an attempt to manage location to control spatial competition is described by Lord in the automobile dealership sector in North Carolina. (13) In this case, manufacturers allocated areas of geographic responsibility to dealers, leading to locational agglomerations which did not compete.

An appreciation by some retailers of the relative importance of individual towns or cities as part of the urban hierarchy has been demonstrated as a clear component of location strategy. Although now quite dated, Schillers examination of retail branch distribution in the UK showed an innate awareness of centre-gravity or 'central place'. (14) Schiller looked at the distribution of the branches of the Mothercare outlets (now a subsidiary of the Storehouse group), demonstrating their use of 'higher centre' locations through their policy of expanding the size of their existing branches in larger centres rather than opening second branches. Dawson and Sparks's construction of a 'post-hierarchical' typology of retail activities complements the work of Schiller by emphasising that different types of retail outlets occupy different sorts of locations because of the underlying consumer needs. (15) For instance, essential shopping (eg. local convenience shops), fun/leisure shopping (eg. large shopping complexes), purposive shopping (eg. variety shops or department stores), time-pressured shopping (eg. petrol station shops), and innovative shopping (eg. catalogue showrooms) all have very different locational requirements which are not necessarily dictated by the strength of centre, but often more by the needs and relative time pressures placed on the customer as a result of social or economic constraints.

A final aspect of the geographic literature on how location is used as part of corporate retail strategy relates to the fact that the catchments which store outlets serve can be the subject of predation or attack by competitor outlets, either through local, spatial marketing of existing stores, or by the onslaught of new outlets encroaching on their catchment areas and resulting in trading impacts. West and von Hohenbalken for example describe how supermarket group Safeway operated a policy of beating competitors into markets in given locations by pre-emptive purchasing of sites or, more aggressively, by locating very closely to competitor supermarkets. (16) This process of locational 'matching', effectively removed one of the key bases for customer differentiation between competitors (locational convenience), forcing them to choose on the basis of other elements of the marketing mix. These aspects of managing existing locations are vital to prevent competitor encroachment given that consumer cognition of urban spatial patterns themselves evolve [see the work of Lee and Schmidt. (17)]

3. THEORISING SPATIAL CHANGE

Beyond the basic components of contagion, avoidance, collusion, hierarchy, consumer targeting, defense and attack described above, the geographers' understanding of the ways in which space is used as part of corporate strategy is still fairly piecemeal. There has recently been a call to develop meaningful models of locational change. (12) But perhaps our only way of achieving this is by recognising that breaking down the artificial barriers of academic disciplines is central to developing our understanding further. As Laulajainen has noted, no discipline has a monopoly of high-level decision-making. (18) A managerially centred outlook would therefore help develop a more coherent view of location as an essential part of the marketing mix.

That is not to say that the call for a more coherent theory has not already been made. A number of authors have stressed the only on-going constant of retailing as being that of change itself, normally from non-ideological perspectives. (19) (20) Others have begun to argue more from a powerful structuralist perspective in order to provide the theoretical underpinnings necessary to understand the factors creating the need for flexibility in retail organizations, especially the 'reflexive' nature of post-modern consumption itself, and the competition between retail companies vying for the consumptive power of individual consumers. (21) (22) (23) Indeed, developments in structuralist and especially Marxist research have come a long way to explain the underlying logic behind change in the capitalist space economy, as demonstrated by the technical work of Sheppard and Barnes and the theoretical paper by Gough. (24) (25) Despite the call for more concerted research on the 'new economic geography of retailing' (Institute of British Geographers 1993

Annual Conference) however, little practical research has yet been done. As an illustration of the practical importance of location management to retail organizations, the next section will examine developments in the UK grocery sector.

4. LOCATIONAL STRATEGY IN THE UK GROCERY SECTOR

Developments in the UK grocery sector are well documented in geographical circles, particularly as a result of the work of Wrigley, who stresses the fact that the high degree of expansion and concentration of the sector within the hands of a few major multiples is a result of the more benign planning regime in the UK compared to, for example, North America. (26) (27) The degree of concentration has increased dramatically since the 1950s, when as Wrigley points out, almost 80% of the market was in the hands of independent grocers and consumer cooperatives, contrasting with the currrent situation when the top handful of grocery chains account for about the same amount.

At the heart of the concentration of the UK grocery sector has been the phenomenal increase in the number and rate of new store openings. This process has had implications far more complex than simply the profusion of retail floorspace; the 'race' for the best remaining locations in the strive for market share has led to very unequal scales of development. Indeed, between the five largest operators - Sainsbury, Tesco, Safeway (now owned by Argyll in the UK), Asda and Gateway - arguably the most telling factor in the success or otherwise of expansion has been the quality of their site location strategies and assessment procedures. Gateway's apparent loss of direction and poor financial position in the mid-1980s led to them selling off over 60 of their outlets to Asda in 1989. Asda's purchase price of £705m has been criticized for being too high, and has led since to an almost total withdrawal from new store development. Financial overextension, coupled with their existing stores being hit by the continued expansion of the stronger retail brands (Sainsbury, Tesco and Safeway), has led to a vicious trading spiral for Asda, a weakness which has been exploited by the remaining stronger operators.

New store openings have formed the primary plank of the location strategies of the 'Big Three' food retailers - Sainsbury, Tesco and Safeway. Between them, they have continued to open 60-70 'superstores' (over 25,000 sq. ft.) annually. Table 1 illustrates the main features of this expansion. The three operators have increased their number of stores by over a third since 1987, mainly by new store development, but also by acquisition (eg. Argyll's acquisition of Safeway in 1987). This net increase of almost 280 stores between them has led to major increases in sales space of at least 50% for all three operators. Moreover, the trend towards larger units (for operational efficiencies as well as increasing their attraction to the customers), has been an essential component of their repositioning; as outlets have moved to largely 'free-standing' sites outside the town centres.

TABLE 1: CHANGES IN UK GROCERY RETAILING 'BIG THREE'

	Increase in Number of Stores, 1987-92						Changes in Store Characteristics			
	1987	1988	1989	1990	1991	1992	Sales Area Increase	Store Size Increase	Av Size sq ft 1992	Sales per sq ft 1992
Sainsbury(1)	283	283	292	291	299	313	+52%	+33%	24,000	£964
Tesco	337	379	374	379	384	396	+49%	+18%	26,000	£678
Safeway(2)	133	176	240	291	310	322	+220%	+142%	20,000	£610

Notes: 1. Sainsbury data excludes Sacacentre hypermarkets.
2. Safeway data includes the acquisition of Presto stores in 1987/88.

Source: Compiled from Mintel. (28)

The result of these major store opening programmes, as Table 2 shows, has been to bring about a joint increase in market share of Sainbury, Tesco and Safeway of 5 percentage points in the three years alone from 1987 to 1990 to 42.3%. The technical expertise and research put into ensuring the quality of these investment strategies has been illustrated by references elsewhere to the Tesco approach to site location as indicative of the importance of new sites. (29) (30) Such efforts, as Table 1 illustrates, have contributed to the creation of some of the highest returns on retail space in the world.

The process of expansion has not yet finished. The top three food retailers still remain relatively 'bullish' in their expansion plans. As the Sainsbury Chairman put it:

> "One third of the British population is still not within reasonable driving time of one of our stores." (31)

TABLE 2: GROWTH IN MARKET SHARE, 1987-90		
Operator	1987 (%)	1990 (%)
Sainsbury	13.9	16.0
Tesco	13.5	15.3
Safeway	9.9	11.0

Source: Compiled from NAMNEWS. (32)

This quotation does, in fact, illustrate a subtle shift underlying the location strategies of the UK food retailing sector's top three organizations, which is towards an increasing emphasis on representation in given regions and towns and cities within them, rather than on simply the absence of modern competition which was the driving force behind so much of the earlier store development. Increasingly, the evidence of expansion programmes demonstrates that these three operators are looking towards measures of company under-representation, rather than simply superstore under-supply, in setting out some of their broader strategic aims. Table 3 demonstrates the regional market shares of the three operators, relative to their average share of the UK market overall.

TABLE 3: REGIONAL REPRESENTATION BY OPERATOR, 1990

Operator	South West	South East	East Anglia	Wales	West Midlands	East Midlands	North West	Yorks & Humber	North	Scotland
Sainsbury	13	145	42	-62	60	-6	4	-25	-72	0
Tesco	3	28	15	60	9	-10	15	15	-74	-62
Safeway	-22	8	-88	-58	2	-64	-46	-4	157	116

Notes: 1. Ratio comparing regional share of each market that the operator has, with the operator's average share of the market in the UK.

Source: Chestertons. (33)

Table 3 also highlights, for example, Sainsbury's under-representation in the North of England and Scotland (both stated areas of expansion priority for the company); which compares with a similar situation for Tesco. By contrast however, Tesco has seemingly decided to concentrate on other regions, possibly because of the strong trading franchise of Safeway in the North and Scotland. The increasing aggression of this strategy has been noted as a new feature of the location strategy of these same three organisations:

> "It is increasingly common for the three major operators to be opening stores close to those of their rivals, increasing the pressure to attract enough shoppers to fill the new supermarkets." (34)

A result of this pressure, noted in the same source, is the increasing emphasis of the management of the superstore chains on places lower down the urban hierarchy: 90% of the 900 stores of over 20,000 sq ft trading at the end of the 1980s were in settlements of over 30,000 people; since 1990 only 75% of the 150 new stores opened were in towns of over 30,000 - and the average size of store in these locations had also reduced to maintain returns. None of the three operators have yet dared to pull out of this'race' as yet, even though it looks likely that the growth in stores is increasingly to prevent erosion of trade by competitors as much as anything else. There are also more stores in the pipeline - the above source notes that capitalised interest - the amount of interest paid on investments held for future developments currently in property'landbanks', indicates in the region of £1.5bn of unfinished supermarkets: 17 Sainsbury's; 31 Tesco's; and 7 Safeway's. The same retailers have all had to resort to huge share issues to raise the finance required to fund the huge ventures (raising almost £1.5bn between them in 1991 by this means).

The race to obtain the prime sites for large superstores has entailed bidding at very high levels for these locations. In 1992 for example, the cost of superstores built by the 'Big Three' averaged £22-23m for store sizes of almost 24,000 sq ft each (see Table 1). The decisions to build these stores were made mainly in the period prior to the relatively recent onslaught of a number of European 'deep discount' chains who make their comparatively austere stores "work like a Swiss watch". (35) The 'up-scaling' of the major operator offerings already described has had a twofold effect of making the market more attractive to potential entrants and raising customers perceptions of price, potentially undermining existing shopping habits. (36)

Commentators now believe that the discount market in the UK may be at least twice as large as was initially thought. (37) It is interesting to note the impact which the discounter location strategies has had on the larger store operators; the nature of their sites ideally being in more convenient locations close to town centres and in poorer areas - by and large areas avoided by the larger firms who are interested in the more mobile and hence more affluent customers. To some degree, the corollary of the two different market positions has led to a spatial polarization of the two types of store. Even so, the impact of discounting has caused the 'Big Three' to carefully reconsider their expansion criteria, a trend likely to continue with the threat of new US-style warehouse clubs opening in the UK (such as Costco and Price Club). Whilst food retail analysts warn about the level of investment in the food sector, some academics such as Neil Wrigley have suggested that such apparent overinvestment in strategic sites by the major operators, may have strategic value in the long-term as 'sunk costs' necessary to help maintain entry barriers to competitors. (38)

Other responses to the onset of saturation have been made by the larger operators. Both Tesco and Sainsbury have undertaken joint venture 'duo' stores (with Marks & Spencer and John Lewis respectively) in an attempt to provide a differentiated offer. They are also beginning to target their locational expansion on major city centres with strong catchments, using small stores (c. 10,000 sq ft) to serve office workers, and tourists as well as local residents, areas which have largely been vacated by food stores previously because of the high rents and distribution problems associated with them. Tesco's 'Metro'concept and Safeway's 'City' format are cases in point. Finally, the larger operators are also being forced to make forays into overseas markets so as to enable growth to continue in the longer term. Sainsbury acquired the Shaw's supermarket chain in New England a few years ago, and Tesco purchased Catteau, a relatively small French chain in northern France in early 1993 to establish a foothold on the Continent.

5. CONCLUSIONS

This initial exploration has illustrated some of the main ways in which retail organizations can use location as an integral component of their corporate strategy. It has emphasised the importance of the pace of new store development, the targeting of specific locations in the urban hierarchy (such as a revival of town centre food stores, or discounters focusing on the poorer geographical pockets), and the creation of new differentiated retail developments (for example, joint store offers) as just some of the responses UK food operators have come up with in the face of market maturity. Such an initial examination of one sector in a single economy cannot profess to be indicative of broader trends, however, it does emphasize the need to begin research along a number of avenues if the issue of how location is used as part of retail strategy is to be developed more fully. In particular, it stresses the importance of viewing store locations as networks of competing brands and the need to look more closely at the management processes associated within different market 'segments', in a variety of retail sectors, at varying levels of development (conception, growth and maturity), and in contrasting international situations. An interdisciplinary approach to these issues - using concepts derived from marketing and the strategic planning literature, as well as geography - will be a vital first step in the process of constructing a perspective of location strategy which is management centered, a perspective which should also prove useful in helping to explain the fundamental changes in the retail landscape which are currently occurring.

6. REFERENCES

1. Davies, R.L., and D.S. Rogers (eds), 1984. Store Location and Store Assessment Research. London: Wiley.

2. Wrigley, N., (ed), 1987a. Store Choice, Store Location and Market Analysis. London: Routledge.

3. Brown, S., 1987. "Institutional Change in Retailing: a geographical interpretation," Progress in Human Geography 11:181-206.

4. Jones K. and J. Simmons, 1990. The Retail Environment. London: Routledge.

5. Ghosh, A. and S.L. McLafferty, 1987. Location Strategies for Retail and Service Firms. Lexington: D.C. Heath.

6. Dawson, J.A., 1979. Marketing Geography. London: Groom Helm.

7. Jain, A.K. and V. Mahajan, 1979. "Evaluating the Competitive Environment in Retailing Using Multiplicative Competitive

Interaction Models," in J. Sheth (ed.), Research in Marketing. Greenwich, JAI Press: 7.

8. Laulajainen, R., 1988. "Chain Store Expansion in National Space," Geografiska Annaler 70B:293-99.

9. Sparks, L., 1990. "Spatial-structural relationships in retail corporate growth: a case study of Kwik Save Group, "Service Industries Journal 10:25-84.

10. Laulajainen, R. and L.E. Gadde, 1986. "Locational Avoidance: A Case Study of Three Swedish Retail Chains," Regional Studies 20(2):131-40.

11. Lord, J.D. and D.B. Wright, 1981. "Competition and Location Strategy in Branch Banking: Spatial Avoidance or Clustering," Urban Geography 2: 189-200.

12. Brown, S., 1992. Retail Location: A Micro-Scale Perspective. Aldershot: Avebury.

13. Lord, J.D., 1992. "Locational Dynamics of Automobile Dealership and Explanations for Spatial Clustering," International Review of Retail, Distribution and Consumer Research 2:238-308.

14. Schiller R., 1981. "A Model of Retail Branch Distribution," Regional Studies 15:15-22.

15. Dawson, J. and L. Sparks, 1986. "Issues for the Planning of Retailing in Scotland," Scottish Planning Law and Practice 18:38-40.

16. West, D.S. and B. von Hohenbalken, 1984. "Spatial Predation in a Canadian Retail Oligopoly," Journal of Regional Science 23:137-54.

17. Lee, Y. and C.G. Schmidt, 1988. "Evolution of Urban Spatial Cognition: Patterns of Change in Guangshou, China," Environment and Planning A 20:339-51.

18. Laulajainen, R., 1988. "The Spatial Dimension of an Acquisition," Economic Geography 64:171-87.

19. Wrigley, N. 1987a op cit.

20. Brown, 1992, op cit.

21. Glennie, P.D. and N. Thrift, 1992. "Modernity, Urbanism and Modern Consumption," Environment and Planning D: Society and Space 10:423-43.

22. Gibbs, D.C. 1992. "'The Main Thing Today is Shopping': Consumption and the Flexibility Debate," Spatial Policy Analysis Working Papers, 15: January, University of Manchester.

23. Ducatel, K. and N. Blomley, 1990. "Rethinking Retail Capital," International Journal of Urban and Regional Research 14(2):207-27.

24. Sheppard, E. and T.J. Barnes, 1990. The Capitalist Space Economy: Geographical Analysis After Ricardo, Marx and Sraffa. London: Unwin Hyman.

25. Gogh, J., 1991. "Structure, System, and Contradiction in the Capitalist Space Ecnonomy," Environment and Planning D: Society and Space 9:433-39.

26. Wrigley, N., 1987b. "Guest Editorial: The Concentration of Capital in UK Grocery Retailing," Environment and Planning A 19:1283-8.

27. Wrigley, N., 1991. "Commentary: Is the 'Golden Age' of British Grocery Retailing at a Watershed?," Environment and Planning A 23:1537-44.

28. Mintel (1993), 'Food Retailing," Retail Intelligence, 1, London.

29. Penny, N.J. and D. Broom, 1987. "The Tesco Approach to Store Location," in N. Wrigley (ed.) Store Choice, Store Location and Market Analysis. London: Routledge.

30. Moore, S. and G. Attewell, 1991. "To Be and Where Not To Be: the Tesco Approach to Locational Analysis," OR Insight 4(1):21-4.

31. See Mintel, 1993 op cit, p. 54.

32. NAMNEWS (1992), ED Newlsetter for National Account Managers, 2(10):9.

33. Chestertons (1991), Food Stores: A Taste of the Future, London.

34. Super Marketing, 1993. "Store Openings: Is the End in Sight?," 22 January: 16.

35. The Economist, 1993. "Europes Discount Dogfight," 8-14 May:81.

36. Duke, R.C. 1993. "European New Entry into UK Grocery Retailing," International Journal of Retail & Distribution Management 21(1):35-39.

37. NAMNEWS, 1993. EC Newsletter for National Account Managers 3(5):15.

38. Wrigley, N., 1993. "Sunk Costs and Corporate Restructuring: British Food Retailing and the Property Crisis," Paper presented to the Institute of British Geographers Annual Conference, Longon: January.

Qualitative Reasoning and Environmental Decision-Making: The Role of Logic Modeling

James K. Lein
Department of Geography
Ohio University
Athens, Ohio 45701-2979

1. INTRODUCTION

When examining the relationship between human populations and their environments, the principal research focus for nearly a century has been centered on the impact of humans on the systems and sub-systems that define planet Earth. While this impact-centered perspective has produced a wide and varied literature documenting the myriad ways humans have contributed to global environmental change, a related and equally critical aspect of the environmental problem has received comparatively limited attention. This somewhat overlooked aspect of the environmental problem describes the decision-making processes that rest at the core of all human-directed environmental change. Therefore, documenting studies explaining the causal relationships that define processes such as acid precipitation, biodiversity loss, and greenhouse warming, although critical to improved scientific understanding, are by themselves, insufficient unless the essential knowledge they reveal can be assembled into a form that can be applied in a decision-making context.

From this perspective, human impact on the environment is the product of a human decision-making process, and the effects, like forced species extinction, are the consequence or by-product of decisions emanating from that process. Making environmental decisions, however, is strongly influenced by the presence of uncertainty with respect to the behavior of the environmental systems involved, and moderated by elements of risk, imprecision, and inexactness introduced through the decision models used to evaluate the dynamics of environmental systems subject to human intervention. Enhancing environmental quality begins not simply by treating the symptoms of degradation, but by fundamentally improving the decision tools used to relate environmental knowledge and information, and the logic systems employed to model human-environmental processes. At issue is the fact that an element of qualitative reasoning pervades all aspects of environmental decision-making that cannot be adequately captured in the current generation of computer-assisted decision models, including GIS. Therefore, a need exists to search out an alternative strategy that can accommodate and combine quantitative analysis with qualitative descriptions of human-environmental processes, or replace quantitative analysis completely where required.

Research in this paper explores the potential application of logic modeling in this role and describes its potential application in connection to the GIS and simulation methods currently employed in automated environmental decision-making.

2. ENVIRONMENTAL DECISION-MAKING IN CONTEXT

Environmental decision-making occurs whenever a decision must be made that affects the present or future quality of the environment. (4) While this definition seems fundamental, deciding on the appropriate course of action given the complex interrelations and interdependencies that characterize environmental systems can introduce an intractable level of uncertainty into the comparatively simple task of making a selection. For instance, siting a high level nuclear waste repository requires consideration of a complex number of physical and human variables extrapolated out over a 50,000 year time horizon. Assigning probabilities to events, developing scenarios of processes, simulating breaching event pathways, and determining consequences across the time horizon relies more on expert judgment than quantified decision-making. (18)

Deciding how to dispose of hazardous waste is clearly an environmental decision, however, there are many other decisions where the environment is not the primary concern, although it will be affected by the decision. (4) Time, as suggested above, also plays a crucial role in complicating the environmental decision. Lein, commenting on the problem of uncertainty in relation to modeling human impact, showed that the speed with which modern technology develops assures that decisions made by society will increasingly risk affecting environmental systems well beyond the planning horizons of present-day socio-political systems. (18) Hence the issue confronting society is not merely the implementation of efficacious, ethical, and optimal decisions, but the challenge of re-aligning the decision-making process to place it into a truly environmental context.

Decision-making takes place within the context of a problem. Recognizing the problem and gaining insight about its features requires more than technology. It also requires an approach that can lead to its solution. However, when confronted with a problem, each individual approaches their solution differently depending on variations in background, experience and psychological conditioning. In general, decision-making styles fall into five broad categories. (2) (7)

(1) The Rational Model - defining decision-making as a structured process where each problem can be reduced to a set of factors that influence the result. Following this classic approach to decision-making, selection moves through a five step sequence: 1) problem definition, 2) description of alternatives, 3) evaluation of alternative, 4) selection , and 5) review of solution.

(2) The Satisficing Model - recognizing that an "optimal" solution may not exist, the satisficing model seeks an "adequate" alternative and relies on feedback to improve the next solution, if possible.

(3) The Organizational Model - following this design, the decision-making process has been simplified to a set of established policies or procedures that are imposed by an organizational unit. The individual in this model avoids uncertainty by adhering to the standard in order to reach a decision.

(4) The Political Model - this model emphasizes the bargaining process. Here, decisions are arrived at through group interaction. Each individual in the group may influence the decision through persuasion or organizational power, but there is no absolutely "best" decision. Rather the goal is to reach a solution that is acceptable to the group.

(5) The Individual Model - at the individual level, decision-making is largely a function of personality, background, experience, and organizational status. These traits or idiosyncrasies help define the "decision style" used or adopted by the decision-maker in selecting a solution that fits individual requirements or the organization.

The models of decision-making summarized above suggest several points of departure that frustrate their application to environmental problems. (4) First, while many decision problems are complex, environmental decisions are generally more so. Decision models typically prescribe simplifying a problem to its essential elements, yet environmental problems are less easily simplified. Consequently, bad environmental decisions can result when decision-makers oversimplify the problem. (20) Secondly, environmental decisions frequently have a greater degree of incertitude. In many cases decision-makers simply do not have complete understanding and are unable to predict or envisage all the possible outcomes and consequences. Also, because the ecological system is a system with inertia, decision actions today may not immediately produce the desired results. Lastly, environmental actions also define environmental risks and hazards that stimulate citizen involvement and political pressures that are not typical of other decision problems. Environmental decision-making is, therefore, greatly influenced by: 1) the model used to characterize the problem, 2) the ability to conceptualize the options and possible outcomes available, and 3) the methods selected to handle uncertainty. The presence of these conditions imply that environmental decisions possess a fundamentally different structure than typical decisions, either corporate or personal.

Simon makes an important distinction between programmed and non-programmed decisions that should help clarify the issues related to the environmental decision-making process and the techniques used to integrate environmental information into a decision model. (23) According to Simon, programmed decisions can be defined as repetitive and routine. In an environmental management application examples of programmed decisions can include organizational decision-making or operational procedures such as permit review, design review, negative declarations, regulatory and compliance checks. Non-programmed decisions are unstructured, unique and call upon the combined exercise of judgment, intelligence and adaptive problem-solving behavior. This class of problem defines the majority of decisions that presently frustrate environmental decision-makers since they not only describe incertitude about the environmental processes involved, but also the imprecision, inexactness, and uncertainty characterizing human-expert judgment.

Keen and Morton have refined the programmed/non-programmed distinction, describing problems as structured, semi-structured, or unstructured. (12) Structured problems are programmable and do not require a decision-maker for implementation. Generally, a well defined and tested set of operating procedures exist to execute decisions of this type. Semi-structured problems explain decisions that emerge from a less well defined set of operating procedures. Unstructured problems are unique and tend to have few operating procedures to guide their implementation. Problems of this type are generally non-recurring or "new", placing a premium on the decision-maker's experience and judgment to obtain an acceptable solution.

3. AUTOMATED DECISION-MAKING

Translating decision-making skills into a form that can be processed by machine has been an active research frontier for over 30 years. While limited success has been achieved with respect to programmed and structured decisions, implementing computer-assisted solutions for non-programmed, semi- and un-structured problems remains elusive since decisions of this type rely heavily on the application of human judgment, and shallow knowledge and experience. Recently, a class of computer programs, referred to as Expert Systems, have been introduced into environmental decision-making arenas. (17) (18) (19)

Although the expert system approach has shown great promise, there are still major limitations that frustrate their widespread application. As suggested by Kuipers, an expert system is often a "shallow model" of its application domain. Genuinely expert performance must also rest on knowledge of "deep models" in which an underlying mechanism, whose state variables may not be directly observable, accounts for the observed facts. (15) An automated decision process, therefore, requires mechanisms for: 1) recognizing situations that call for actions, 2) formulating problems, 3) designing actions and setting goals, and 4) evaluating alternatives and making selections.

The current generation of automated decision tools employ the decision support paradigm that was first introduced in the early 1970s. Following this model, the computer system provides information in a given domain by means of analytical decision models and access to databases in order to support a decision-maker in making decisions effectively in complex and ill-structured (non-programmable) tasks. (14) Problem structure is a key concept in this design since the process relies on human guidance to reconcile the unstructured aspects of the problem. (3). If the decision support system is to function, the decision-maker must possess key traits to bridge the analytic gap: perception, judgment, decision power, analytic power, and adaptation. (14) The decision support system alone reflects a store of information and knowledge about the domain in the form of data variables, models, reports, and graphics, but it cannot draw conclusions or explain how a decision was reached. To accomplish the later two goals requires expertise and the ability to explain model behavior, determine the logical structure of the model itself, and the causal relationships among variables. Such faculties require embedding in the system the capacity to perform qualitative reasoning tasks.

4. QUALITATIVE REASONING AND SIMULATION

Change is a ubiquitous characteristic of the environmental system. (13) Years of scientific investigation have produced a rich and diverse physics that provides many answers, and concepts have been derived, such as state, cause, momentum, feedback etc., to characterize change. Yet these concepts, although commonly defined quantitatively are qualitative and can be intuitively understood. (13) Furthermore, while the behavior of an environmental process can be described by the exact values of its variables, such a description fails to provide much insight into how the system functions. As deKleer and Brown point out, the insightful concepts and distinctions are usually qualitative, but embedded within the much more complex framework established by continuous real-valued variables and differential equations. (13) Forbus, recognizing that much of formal physics consists of characterization of process by differential equations that describe how the parameters change over time, concluded that the notion of process is richer and more structured than what can be explained by these equations. (8) This observation carries the implication that people often reach conclusions about physical processes based on very little information. To design decision support systems that can function when confronted with semi- or un-structured problems requires identifying this "core" knowledge that underlies physical intuition concerning the environmental processes involved. Once identified, this knowledge can be used to develop decision models that draw on physical intuition.

Environmental systems, however, are characterized by many sources of uncontrolled variation. Therefore, it is often difficult to know the initial state of the system, the role played by external factors, and how variation is introduced to and through the system. When a model is constructed, it reflects the builder's understanding of the process(es) involved, and as such it can be regarded as a scientific hypothesis. (22) The hypothesis is evaluated by comparing the behavior of the modeled system with that of the real world. When decision-makers reason about a process they too apply a model to the situation, but rather than a rigorous quantitative description of process and causality, the model is qualitative in nature and what can be referred to as a qualitative causal calculus. (13) Future design of Expert-Geographic Information Systems and Intelligent Spatial Decision Support Systems, will therefore need to develop decision models based on this qualitative calculus in order to substantively improve their abilities to address non-programmed problems.

Presently, the expert systems have been the principal vehicle for automated problem-solving that requires extensive human expertise. However, a recognized failing of these systems is their extremely narrow range of expertise coupled with their inability to recognize when a problem posed to them is outside their range of knowledge; they lack common sense. (13) To produce decision models that can apply common sense physical reasoning, one must understand how to reason qualitatively about processes, when they will occur, their effects, and when they will stop. (8)

Constructing a qualitative model begins with a scenario that identifies the entities in the system one is interested in modeling, and specifies their initial conditions. (6) The entities in the scenario become part of the initial model, where the scenario in this context is essentially a sequence of processes and events that can be characterized by a logic system connecting the specified parameters. (18) The product is a self-explanatory simulation that consists of a tightly integrated set of qualitative and numerical representations. Such a simulation has three components : 1) a set of defining procedures that specify for each qualitative state how to update its numerical parameters over time, 2) a set of state transition parameters that detect qualitative changes in state, and 3) an explanation facility that uses information from the scenario to provide causal accounts and characterizes possible behaviors. (9) One important aspect of a qualitative simulation is that the behavioral prediction can branch, corresponding to qualitatively distinct futures that cannot be discriminated by the available information. (10)

5. DEVELOPING LOGIC MODELS : A SIMPLE BEGINNING

Ideas can have strong intuitive appeal, yet not affect decision-making because they lack an explicit operational formulation. (21) Cumulative environmental change is one such idea and exemplifies a critical non-programmed environmental problem presently confounding decision-makers. At issue is the observation that individually insignificant actions can produce major change through the accumulation of effects. The problem is that cumulative impacts are not always obvious, nor easily attributed to a particular disturbance. (11) If a process orientation is taken to help clarify the interpretation of cumulative change, the concept assumes three principal dimensions: 1) sources of change, 2) pathways of accumulation, and 3) impact accumulations. Embedded in this description, however, is a typology of qualitative variables, and a growing realization that the process of cumulative changes may be non-linear. Not surprisingly, attempts to explain the process of cumulative change utilizing traditional modeling techniques have been largely unsuccessful. Replacing traditional approaches to cumulative change assessment with a logic programming approach based on controlled deduction may hold promise for this and similar examples.

Formulating a qualitative process model based on controlled deduction requires:

(1) Developing a "picture of the world" characterizing the region of interest as a series of object relations contained within frames that explain the salient features of the region.

(2) Constructing a set of problem description rules to control dialogue with the model user.

(3) Structuring a set of modeling rules that capture the behavior of the system (region) and organize the object frames according to known physical principles and expert judgment.

(4) Placing a scenario into the frame-net and examining the resulting changes in the object-relations that form the frame-net.

Several qualitative process models and compilers have been constructed according to the outline given above to evaluate the utility of qualitative reasoning as a modeling device; the EL Prototype (22), Q3 (1), QP (8), and QSIM (16). However, none have been applied to environmental situations.

Qualitative modeling building, whether focused on the problem of cumulative environmental change, or any other non-programmed environmental decision must build on the concepts taken from qualitative processes theory (QPT). (8) According to QPT, there are two fundamental types of descriptions: Individual Views, which represent objects or sets of objects viewed in a particular way, and Processes, which represent active changes taking place; where all change is considered to emanate from processes. From this beginning, the model building process leading to a simulation can be summarized as follows. (16)

(1) Each individual view and process checks various conditions about the "world" to determine whether it should have one or more active instances.

(2) The active view and process instances are grouped into sets of mutually consistent elements, termed view-process structures.

(3) Each view or process in a view-process structure contributes fragments, called direct or indirect influences, to the constraint model.

(4) A closed world assumption is applied within each view-process structure, making it possible to determine the set of all influences, direct or indirect, applying to any given quantity, allowing the influences to be translated into constraints.

(5) The resulting constraint model can then be simulated.

Refining this approach to the point where "real world" systems can be simulated rests on three main issues: 1) developing more powerful mathematical methods to enhance selection of the appropriate level of qualitative description, 2) developing techniques for the hierarchical decomposition of complex models, and 3) developing strategies to control the incremental application of quantitative knowledge. (16) Once progress in these three key areas has been made, a tractable methodology will be available to substantively improve the GIS and Expert System approaches to ill-structured environmental problems.

6. CONCLUSIONS

Environmental decision-making is unique to the extent that the subjects and processes involved are often ill-defined, poorly understood, difficult to measure, and uncertain in their behavior. The use of GIS and more recently the application of Expert Systems has been advocated as viable approaches to improve the modeling of environmental relations. Yet both techniques fail when confronted with incomplete knowledge and unstructured problems. In this paper the techniques of qualitative reasoning and logic modeling were reviewed within the context of environmental decision-making. This review has shown that qualitative reasoning methods provide greater expressive power for states of incomplete or inexact knowledge than traditional modeling approaches. It was further shown that even with incomplete knowledge there is generally enough information available in a qualitative description to make meaningful prediction feasible. Based on these findings, a new generation of decision models can be constructed that offer the potential to greatly enhance the study of environmental change.

7. REFERENCES

1. Berleant, D. and B. Kuipers, 1992. "Qualitative- Numeric Simulation With Q3." IN: Recent Advances in Qualitative Physics. B. Falting (ed). The MIT Press, pp. 3-16. Cambridge, MA

2. Bidgoli, H., 1989. Decision Support Systems. St. Paul, MN: West Pub. Co.

3. Bonczek, R., 1981. Foundations of Decision Support Systems. New York: Academic Press.

4. Chechile, R., 1991."Introduction to Environmental Decision Making." Environmental Decision Making: A Multidisciplinary Perspective. R. Chechile (ed). New York: Van Nostrand Reinhold, pp. 2-22.

5. Cocklin, C., 1992. "Notes on Cumulative Environmental Change I: Concepts and Issues." Journal of Environmental Management 35: 31-49.

6. Crawford, J., 1992. "QPC: A Compiler from Physical Models into Qualitative Differential Equations." IN: Recent Advances in Qualitative Physics. B. Falting (ed). The MIT Press. Cambridge, MA.: pp. 17-32.

7. Davis, M., 1988. Applied Decision Support. New Jersey: Prentice-Hall.

8. Forbus, K., 1984. "Qualitative Process Theory." IN: Qualitative Reasoning About Physical Systems. D. Bobrow (ed). The MIT Press pp. 85-168. Cambridge, MA.

9. Forbus, K. and B. Falkenhainer, 1992. "Self-Explanatory Simulations: Integrating Qualitative and Quantitative Knowledge." IN: Recent Advances in Qualitative Physics. B. Falting (ed). MIT Press pp. 49-66.

Cambridge, MA.

10. Fouche, P. and B. Kuipers, 1992. "An Assessment of Current Qualitative Simulation Techniques." IN: Recent Advances in Qualitative Physics. B. Faltings (ed). The MIT Press pp. 263-277. Cambridge, MA.

11. Johnston, C., 1988. "Geographic Information Systems for Cumulative Impact Assessment." Photogrammetric Engineering and Remote Sensing 54(11): 1609-1615.

12. Keen, P. and S. Morton, 1978. Decision Support Systems, An Organizational Perspective, Reading, MA.: Addison-Wesley.

13. deKleer, J. and J. Brown, 1984. "A Qualitative Physics Based on Confluences," IN: Qualitative Reasoning About Physical Systems. D. Bobrow (ed). Cambridge, MA.: The MIT Press pp. 7-84.

14. Klein, M. and L. Methlie, 1990. Expert Systems: A Decision Support Approach. Reading, MA.: Addison-Wesley.

15. Kuipers, B., 1986. "Qualitative Simulation." Artificial Intelligence 29: 289-338.

16. Kuipers, B., 1989. "Qualitative Reasoning: Modeling and Simulation With Incomplete Knowledge." Automatica 25(4): 571-597.

17. Lein, J., 1989. " An Expert System Approach to Environmental Impact Assessment." International Journal of Environmental Studies 33: 13-27.

18. Lein, J., 1992. "On The Application of Expert Systems in Environmental Performance Assessment." Journal of Environmental Systems 21(2): 167-183.

19. Lein, J., 1993. "Applying Expert System Technology to Carrying Capacity Assessment." Journal of Environmental Management 37: 63-84.

20. Moomaw, W., 1989. "Our Record With The Environmental Crystal Ball." EPA Journal May/June: 37-39.

21. Preston, E. and B. Bedford, 1988. "Evaluating Cumulative Effects on Wetland Functions." Environmental Management 12(5): 565-583.

22. Robertson, D., 1991. Eco-Logic: Logic Based Approaches to Ecological Modelling. The MIT Press. Cambridge, MA.

23. Simon, H., 1960. The New Science of Management Decision. New York: Harper & Row.

ANALYSIS OF NETWORK STRUCTURE FOR U.S. DOMESTIC AIRLINE NETWORKS

Jay Lee, Alec Boros
Department of Geography
Kent State University
Kent, Ohio 44242-0001

Shih-Lung Shaw
Department of Geography
Florida Atlantic University
Boca Raton, Florida 33431-0091

1. INTRODUCTION

Historically, an important issue for air transportation policy-makers has been how well places in the United States are served by air carriers. Prior to 1978, such an issue was one of the factors in regulating air transportation. Before deregulation, service patterns of individual airlines were designed to provide intensive coverage within federally-limited geographic boundaries. However, the 1979 Airline Deregulation Act (ADA) ended 40 years of such airline regulation. A study conducted by Brenner et al. in 1985 (4) completed six years after deregulation, concluded that it was impossible to adequately judge the effects of the ADA both in terms of the level of service and the overall concentration of airline networks in the United States. However, Brenner concluded that deregulation has changed the patterns of air carriers such that individual airline coverage has been replaced by national overlapping networks through the use of hub and spokes. (4) [Review and description of the hub and spoke system have appeared extensively in recent literature. (14, 15, 17)]

It is now 14 years after deregulation and we believe it is important to determine the effect of the hub and spoke system on the level of concentration in the U.S. airline industry. A number of researchers have previously focused on centrality, connectivity and diameter to analyze transport networks. (8, 16, 18) However, these analyses failed to uncover the spatial patterns of geographical phenomena in the transportation network. We therefore propose that even though deregulation had created a hub and spoke system which seemingly created a more concentrated pattern which focused on the hubs, the real effects of deregulation on concentration may not be adequately judged based on examining increased connectivity alone. To further our understanding to this end, we apply the concept of spatial autocorrelation to analyzing attributes in transportation networks. Network Spatial Autocorrelation (NSAC) is used in this paper to study the influence of hubs in the U.S. airline industry and to determine the level of concentration.

2. DEFINITION OF SPATIAL AUTOCORRELATION

Spatial Autocorrelation among spatial objects has received increasing attention in recent geographic literature. It has been applied in a number of studies involving geographic phenomenon (1, 7, 13) via twopopular indices: Geary's ratio and Moran's index. (10, 11) However, few have focused specifically on NSAC. Ebdon provides an introductory discussion of NSAC. (9) Griffith used NSAC to analyze grain distribution in the Canadian Prairie Provinces, and most recently Black used NSAC to examine migration flows. (3, 12)

The spatial autocorrelation coefficient (SAC) is generally used to study the spatial patterns of attribute values over a set of spatial objects. SAC is concerned with the spatial patterns shown by the relative location of points, areas, or networks as well as the spatial dependency shown among these objects. In this sense, SAC deals with both spatial and attribute data. (5)

Specifically, SAC measures whether two dimensional geographic patterns are systematic or random. If a systematic pattern can be identified in the attribute values, the examined attribute is said to be spatially autocorrelated. Although care should be given in labeling the results of SAC, three general spatial arrangements can be discerned: clustered, disperse and random. When adjacent spatial objects are more similar than distant objects the resulting pattern is said to be clustered. Alternatively, a disperse pattern is when adjacent objects are not similar or when they are arranged in a uniform manner. Finally, a random pattern refers to an arrangement where no systematic pattern can be detected.

Techniques are available for using SAC to test nominal, ordinal and interval data for both point patterns, networks and areas. (10, 11) There are three commonly used measures of SAC depending on the measurement scale of the spatial variable and whether the data set is concerned with contiguous areas or point and network patterns. If the focus is on contiguous areas with a binary variable then SAC is measured with the join count statistic. If the analysis concerns network or point patterns and/or if interval, ordinal, or ratio data is used then SAC can best be measured by the equations described in the next section.

2.1 SPATIAL AUTOCORRELATION COEFFICIENT

The main process of our analysis involves measuring the strength of SAC for attribute distribution among spatial objects and testing the randomization assumption for the statistical significance of the spatial autocorrelation. The spatial objects in our airline networks are the flight lines among airports and the attribute is the flight frequencies along these flight lines. To define spatial adjacency for our study, we assign a weight of 1 between two flight lines if they share a common airport; otherwise, a weight of 0 is given.

We assume that an index value for SAC is only a descriptive summary of spatial pattern but not the strength of it. Testing the statistical significance of SAC is required if meaningful comparisons are to be conducted. In order to test the significance of SAC it is essential to know how probable it is that the observed pattern could be by chance. To do this we form a hypothesis H0 which states that the observed pattern has been generated by random. There are two possible forms of the null hypothesis in a significance test: normality (free sampling) and randomization (non-free sampling) assumptions. The normality test assumes that all attribute values come from a population of a normal distribution. With the randomization test, however, the particular spatial arrangement of attribute values associated with spatial objects is considered in relation to all the ways in which the values could possibly be arranged within the study region. This takes into account only the situation within the study region and no reference is made to outside factors. In general, it is best to use the non-free sampling significance test when there is any doubt whether the outside region distribution will apply to the study region. (9) The calculated value of NSAC can easily be converted into a z-score for a standard Z-test by using the following equation:

$$Z = 2\frac{NSAC - E(NSAC)}{Var(NSAC)1/2}0 \qquad (1)$$

Calculated values of SAC can be tested for their statistical significance by comparing the above value with critical values of Z at given significance levels.

2.2 ANALYSIS OF NETWORK SPATIAL AUTOCORRELATION

Similar to SAC, NSAC can be used as a descriptive technique to examine a spatial pattern across the arcs of a network thus providing a single measure to reveal tendencies toward clustering, dispersion or randomness. (3) "Spatial Autocorrelation arises because the properties of a place are not simply a function of other properties of that place, but are affected by its ties to other places and the flows of influence that accompany these ties." (2) This implies that NSAC does not arise instantaneously, but the process develops over time and that the places that are not contiguous but connected in some other way may influence the overall spatial pattern. SAC deals with how a variable associated with locations are influenced by contiguous locations. NSAC, however, is conceptually different, as Black explains, "Network autocorrelation concerns with the dependence of variable values on given links to such values on other links to which it is connected in a network context." (3) Despite its applicability, NSAC has been used relatively infrequently in examining the spatial pattern of networks. The use of SAC techniques in network applications has been sparse because the results of SAC provide a single value for a particular geographic pattern. Therefore, the emphasis is taken off individual geographical units and placed on the overall spatial pattern. Subsequently, this paper will develop a technique whereby the overall NSAC can be calculated while allowing for the influence of individual links to be determined.

As previously mentioned, two indices are available for NSAC: Geary's c and Moran's i (see 10 and 11 for detailed discussions). We used Moran's i in this study as Cliff and Ord (6) stated, "Moran's i may be more preferable in many applications because it means that the calculated i will be less affected by deviations from a normal distribution then c." The specific coefficient of Moran's index, i is given as:

$$I = \frac{{}^{\wedge}F_i {}^{\wedge}F_j\, w_{ij}\, c_{ij}}{S2\ {}^{\wedge}F_i {}^{\wedge}F_j\, w_{ij}} \qquad (2)$$

where

$c_{ij} = (z_i - z_{mean})(z_j - z_{mean})$
$S2 = {}^{\wedge}F_i\, (z_i - z_{mean})2 \,/\, n$
$w_{ij} = 1$ if i and j are adjacent, 0 otherwise.

NSAC can be measured by examining the influence of links. Thus attributes of the links are being measured, examples of which include traffic flows, accidents, or tonnage. (3) Low NSAC scores would imply that local factors influence the overall pattern of the network. Likewise, a high NSAC score implies that regional factors are influencing the network pattern. (10)

3. UNITED STATES AIRLINE NETWORK DATA

3.1 AIRLINE NETWORK DATA

The data examined in this paper are based on the number of daily domestic flights for nine major airlines in 1992. These airlines are: American, American West, Continental, Delta, Northwest, Southwest, TWA, United, and U.S. Air. The number of airports that were included in our airline network data was limited to the 100 largest in terms of total U.S. flight volume. For each flight line between two airports, we recorded the origin and destination airports, and the frequency of the flights. Each of the nine airline networks was coded by flight lines. Only those flights between the 100 largest airports were included. Due to the different configurations of airline networks by each airline company, not all 100 largest airports are covered by all airline networks.

3.2 NETWORK SPATIAL AUTOCORRELATION FOR AIRLINE FLIGHT DISTRIBUTIONS

Based on the principles of connectivity developed by Kansky (16), we would expect with the advent of the hub and spoke that the present airline pattern would be highly concentrated. By using NSAC we can shed light on new insights to these spatial patterns. Recall that strong positive NSAC implies a more regional scale of variation. Otherwise, it implies a stronger localized scale of variation. Table 1 summaries the relationship between coefficient values and associated spatial patterns.

TABLE 1 INTERPRETATION OF CALCULATED VALUES OF NSAC

Conceptual patterns	Moran's i
Similar, regionalized smooth, clustered	i > 0*
Independent, uncorrelated, random	i < 0*
Dissimilar, contrasting, checkboard	i < 0*

*The precise expectation is -1/(n-1) rather than 0.
Source: (10, page 16)

Table 2 shows the overall NSAC scores for each airline by using Moran's i. The index values were computed based on flight frequencies of flight lines. Additionally, levels of significance at .01, .05, and .10, were calculated for each coefficient value to indicate their statistical significance. According to Moran's i in Table 2, all airlines display a significant clustering in their flight frequency distribution as all index values are positive and significant at the .05 level or higher. Specifically, Trans World Airline displays the strongest clustering in flight frequencies, followed by Northwest Airline, Continental Airline, and American Airlines. To the other extreme, the American West Airline is the least clustered among the nine airlines examined, followed by US Airline, United Airline, and Southwest Airline.

Two regional airlines, American West and Southwest, show signs of weak clustering, indicating a weaker hubbing structure. In turn, larger airlines with stronger clusterness indicate stronger hubbing structures in their flight frequency distribution. All airline networks shown in Table 2 exhibit positive NSAC which tend to suggest regional influences in the spatial pattern. Therefore, the influence of hubs may be an important causative factor in determining the number of flights to and from a given place and consequently the resulting overall pattern. We next turn to removing individual links and the recalculation of NSAC to determine each links' influence on the overall pattern.

TABLE 2
COMPARISON OF TWO NSAC INDICES WITH NO LINKS REMOVED

Airline	Moran's i
Trans World	.1829***
Northwest	.1005***
Continental	.1004***
American	.0697***
Delta	.0463***
Southwest	.0439**
United	.0400***
US Air	.0363***
American West	.0181**

*, **, *** are significant at the .10, .05 and .01 levels respectively.

3.3 AIRPORTS IN AIRLINE NETWORKS

In order to evaluate the importance of individual airports within an airline network, we devise the following algorithm for computing the differences in NSAC with and without an evaluated airport for a given airline network:

```
    Compute NSACall with the entire network
FOR each airport to be evaluated,
            Remove the airport and all flight lines associated with it
            Re-compute NSACind with the remaining airports and flight
             lines
            Compute the difference of (NSACall - NSACind)
    NEXT airport.
```

Computed in this fashion, we derived the difference in NSAC with and without each evaluated airport. For a given airport, if NSAC decreases when the airport is removed from the network, it means that the entire network becomes less clustered without it. Alternatively, if NSAC increases when this airport is removed, it implies that the inclusion of the airport makes the entire flight distribution less disperse. Larger increases/decreases in NSAC when airports were removed from the network give clues to the importance of the airports in their respective networks.

Tables 3, 4, 5, and 6 and their corresponding maps are for four selected airlines in our data set. These four were selected because they show the most obvious hubbing structure topologically. The maps for each airline shows the pattern of flights among the included airports. The tables that correspond to each map give the six most important airports in terms of largest difference (from NSAC values calculated in Table 2), according to the aforementioned algorithm.

Hubbing structures in airline networks can be complex depending on the locations of airports and the topology of flight frequencies. By topology alone, a hub is an airport where it connects to a larger number of flight lines than those of non-hub airports. However, our approach of using NSAC examines another perspective in that a hub has to be an airport to which the most flight frequency is linked, not just the number of links. That is, an airport may not be a hub although it has many links connecting to it if all those links have relatively low flight frequencies. Alternatively, an airport may be a hub even though there are only some flight lines connecting it to all those carries with larger flight frequencies. In this way, topological connectivity as well as actual flight frequencies are incorporated in the evaluation of hubbing structures.

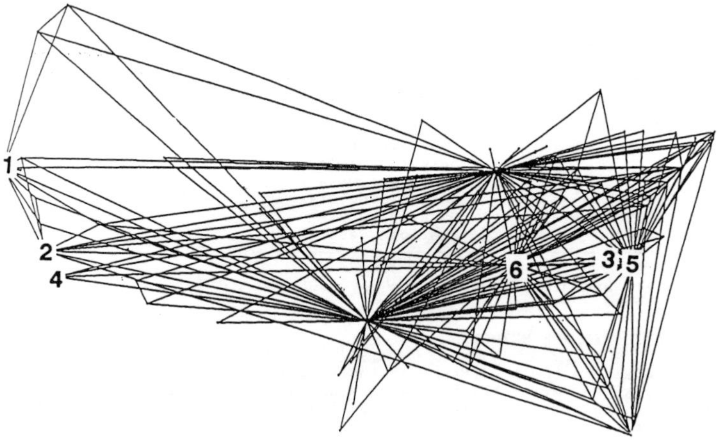

TABLE 3: AMERICAN AIRLINES: 6 AIRPORTS PRODUCING THE LARGEST CHANGE IN NSAC AFTER LINK REMOVAL.

SFO[1]	LAX[2]	GSO[3]	SAN[4]	RDU[5]	BNA[6]
.0234	.0227	.0035	-.0031	-.0170	-.0233

1,2,3,4,5,6 are labled on the map below.
All values are significant at the .01 level using Moran's i.

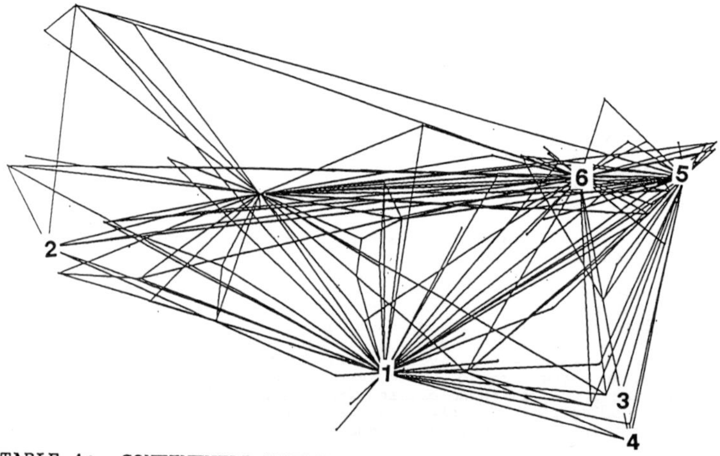

TABLE 4: CONTINENTAL AIRLINES: 6 AIRPORTS PRODUCING THE LARGEST CHANGE IN NSAC AFTER LINK REMOVAL.

HOU[1]	LAX[2]	MLB[3]	MIA[4]	JFK[5]	CLE[6]
.0234	.0166	.0072	-.0121	-.0183	-.0233

1,2,3,4,5,6 are labled on the map below.
All values are significant at the .01 level using Moran's i.

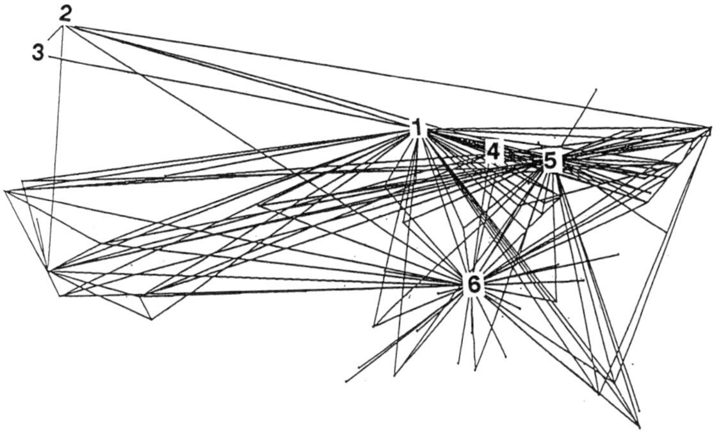

TABLE 5: NORTHWEST AIRLINES: 6 AIRPORTS PRODUCING THE LARGEST CHANGE IN NSAC AFTER LINK REMOVAL.

MSP[1]	SEA[2]	PDX[3]	MKE[4]	DTW[5]	MEM[6]
.0534	.0204	.0125	-.0107	-.0308	-.0564

1,2,3,4,5,6 are labled on the map below.
All values are significant at the .01 level using Moran's i.

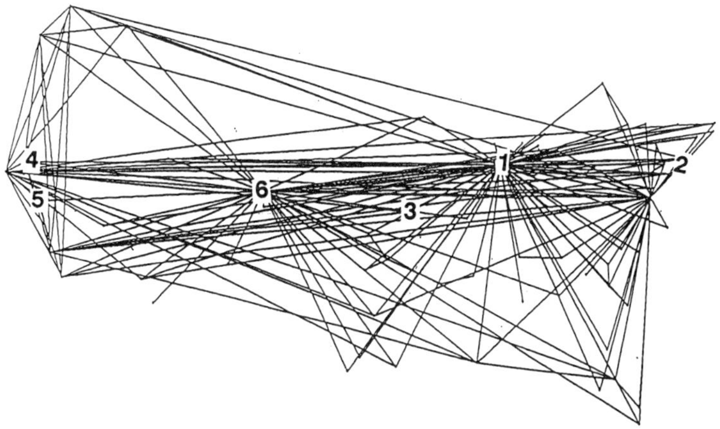

TABLE 6: UNITED AIRLINES: 6 AIRPORTS PRODUCING THE LARGEST CHANGE IN NSAC AFTER LINK REMOVAL.

ORD[1]	JFK[2]	MCI[3]	SMF[4]	FAT[5]	DEN[6]
.0154	.0052	.0038	-.0069	-.0072	-.0150

1,2,3,4,5,6 are labled on the map below.
All values are significant at the .01 level using Moran's i.

Table 3 shows the network and highest and lowest NSAC differences for American Airlines after individual airports were removed one at a time. Judging from the topological pattern, American Airlines has four hubs. Only two of these hubs were included as being the six most significant in terms of differences from the NSAC with no links deleted. Both of the hubs identified by this approach caused decreased NSAC when removed, suggesting them to be important hubs. Note also that when hubs 1 (San Francisco) and 2 (Los Angles), were removed the NSAC increased. This suggests that the pattern became more concentrated and that these West Coast cities have a profound influence on the overall spatial pattern. In other words, although these cities are not used as hubs by American, the number of flights along the links are significantly high.

Table 4 shows the network with the highest and the lowest NSAC scores for Continental Airline after individual airports were removed one at a time. Continental Airline also appears to have four hubs, three of which are identified by our analysis as being the most significant in the recalculated NSAC scores. The two largest negative changes are associated with the Cleveland and New York (New York) hubs. By removing the links of these two airports the resulting pattern was more random. However the greatest positive change also occurred at the Houston hub location and resulted in a more spatially concentrated pattern. This occurs because most of the Continental flights run east-west.

Table 5 was calculated in the same manner. The removal of links for each individual airport clearly identified all three Northwest Airline hubs as having the greatest effect on NSAC. Similar to the patterns described above, Detroit and Memphis caused negative changes, while Minneapolis-St. Paul created the greatest positive change. Clearly other factors are influencing the spatial pattern. For example, Minneapolis-St. Paul may be a collection point for all flights going overseas.

4. CONCLUSION REMARKS

In summary this paper demonstrated that NSAC is an efficient tool for identifying how location influences the overall spatial pattern of an attribute. In our example analyzing airline networks we have shown that connectivity alone cannot fully describe the pattern. Hubbing structures become more complicate if one includes the attributes of flight frequencies. It is not surprising that the topological hubs are different from those hubs identified by NSAC since NSAC considered more than just the topology.

5. REFERENCES

1. Bannister, G. 1975. "Population Change in Southern Ontario." Annals, Association of American Geographers 65: 177-88.

2. Berry, B.J.L., 1971. "Problems of Data Organization and Analytical Methods in Geography." Journal of the American Statistical Association 66: 510-23.

3. Black, W.R., 1992. "Network Autocorrelation in Transport Network and Flow Systems." Geographical Analysis 24: 207-22.

4. Brenner, M., J. Leet, and E. Schott, 1985. Airline Deregulation. Westport, Connecticut: Eno Foundation.

5. Cliff, A. et al., 1975. Elements of Spatial Structure: A Quantitative Approach. Cambridge: Cambridge University Press.

6. Cliff, A. and J. Ord, 1981. "The Effects of Spatial Autocorrelation on Geographic Modelling." In: Future Trends in Geomathematics. R. Craig and M. Labovitz (eds). London: Pion Press pp. 108-37.

7. Dacey, M.F., 1966. "A County-Seat Model for the Areal Pattern of an Urban System." Geographical Review 56: 527-42.

8. Davis, P., 1974. Data Description and Presentation. Oxford: Oxford University Press.

9. Ebdon, D. 1985. Statistics in Geography. Oxford: Blackwell.

10. Goodchild, M.F., 1987. Spatial Autocorrelation. Concepts and Techniques in Modern Geography, No. 48.

11. Griffith, D., 1987. Spatial Autocorrelation: A Primer. Washington, D.C.: The Association of American Geographers.

12. Griffith, D., 1982. "Dynamic Characteristics of Spatial Economic Systems." Economic Geography 58: 177-96.

13. Haining, R., 1978. "A Spatial Model for High Plains Agriculture." Annals, Association of American Geographers 68: 493-504.

14. Houston, J.H. and R.V. Butler, 1991. "The Location of Airline Hubs." Southern Economic Journal 57,4: 975-81.

15. Irwin, M.D. and J.D. Kasarda, 1991. "Air Passenger Linkages and Employment Growth in U.S. Metropolitan Areas." American Sociological Review 56: 524-37.

16. Kansky, K.J., 1963. Structure of Transportation Networks. University of Chicago Research Paper No. 84.

17. O'Kelly, M.E., 1986. "Activity Levels at Hub Facilities in Interacting Networks." Geographical Analysis 18: 343-56.

18. Taaffe, E.J., R.L. Morrill and P.R. Gould, 1963. "Transport Expansion in Underdeveloped Countries: A Comparative Analysis." Geographical Review 53: 503-29.

ABSTRACTS FOR

THE SIXTEENTH ANNUAL

APPLIED GEOGRAPHY CONFERENCE

October 13 - 16, 1993

Ryerson Polytechnical Institute

The Westbury Hotel

Toronto, Ontario
Canada

The abstracts published in this volume are organized alphabetically by author's last name. Only Conference abstracts received by June 15, 1993 were edited and published.

David P. Amborski
School of Urban & Regional Planning
Ryerson Polytechnic University
Toronto, Canada

CONTROLLING DEVELOPMENT IN THE GREATER TORONTO AREA

Metropolitan Toronto has been in existence for almost forty years. It was one of the first Metropolitan governments in North American and has been hailed as an example for both government structure and planning. As time passes, urban growth continues and the complexities of urban areas increase, the viability of Metro Toronto in terms of government form and planning unit must be questioned.

The purpose of this paper is to examine planning in the Metropolitan Toronto and greater area in the context of what might be termed growth management. The concept of planning on a larger than city scale has been applied in Toronto for more than forty years. It began with the formation of Metropolitan government in 1954. It could be argued that part of the rationale for this innovative concept was some of the objectives that are identified in growth management in other jurisdiction today.

Over time, the provincial government undertook several planning initiatives in an effort to plan for growth in the Toronto centered area as concerns arose regarding growth in the region. These initiatives included the "Toronto Centered Regional Plan" (TCR), in 1971, and the associated "Central Ontario Lakeshore Urban Complex Study" (COLUC), in 1974. These earlier initiatives will be examined as they provide the context for provincial planning initiatives in southern Ontario at a scale larger than the Metro level prior to the current Greater Toronto Area (GTA) initiative which began in 1988.

Prior to discussing the details and status of the GTA study and the earlier government initiatives of provincial planning at a large scale, the context of planning in the Province of Ontario will be described with specific reference to those aspects which lend themselves to growth management controls as practiced in other jurisdiction.

Sonia I. Arbona
Department of Geography
and Earth Science
Shippensburg University

John M. Hunter
Department of Geography
Community Health Science
Michigan State University

LAND USE AND WATER QUALITY IN PUERTO RICO

Many categories of land use contribute to point and non-point sources of water pollution. This study addresses the relationship between geographical patterns of land use and the quality of both surface waters and groundwater. A total of 33 sampling stations distributed among 33 representative watersheds were chosen. Major land use categories were in spatial association with the sampling stations. Laboratory analyses for 15 water quality parameters were conducted three times over a six-month period. The results of these analyses showed considerable variation among different land uses but the following associations were recognized: (1) heavy metals contamination with agricultural land use; (2) semi-volatile organic compounds with industrial and residential land use; and (3) nitrogenous contaminants with residential, industrial and agricultural uses, and sewage treatment plants. Water quality standards were breached at most sample stations. The presence of pollutants in groundwater samples, even in small concentrations, threatens the potability of the groundwater reservoir. Monitoring of water quality must be made responsive to changes in land use patterns in order to maintain standards consonant with public health.

Alan M. Baker
Department of Geography
University of Toronto
Toronto, Ontario M5S 1A1 Canada

SEASONAL DEPRESSION AND HUMAN EXPOSURE TO LIGHT

Extensive biological research on mammals over the past 35 years has shown that animals respond to the daily cycle of environmental light and darkness. Their response is indicated by the pineal gland's secretion of the hormone melatonin during darkness, and its suppression during exposure to light. The same daily rhythm persists for a time, even during constant darkness, driven by an internal biological clock.

Until 1980 it had been assumed that humans had evolved in a different way, so that they did not respond to environmental light cycles. Then researchers at the National Institute of Mental Health in Bethesda, Maryland reported the discovery that the response of normal humans to light resembles that of other mammals. They found that exposing the eyes to bright artificial light produced a response, melatonin suppression, similar to that in animals. The light intensity melatonin suppression, similar to that in animals. The light intensity (illuminance) required to produce a response in humans is well below that of typical outdoor environmental light, but well above that of typical indoor environmental light.

Two years later these researchers reported the use of bright artificial light in the successful treatment of a manic-depressive patient with a seasonal mood cycle. During the subsequent decade a flood of research and clinical testing has shown similar results in a majority of patients suffering from the same illness, seasonal affective disorder (SAD). Epidemiological studies have found that most SAD patients become depressed in the winter and recover in the summer, that the prevalence of SAD increases with latitude, and that about 5% of the general population likely has SAD at a latitude of 40 degrees.

Typical treatment of SAD, under medical supervision, involves the use of a portable device, costing a few hundred dollars, in the patient's home or work place. The device exposes the eyes to bright light for an hour or two each day, and typically produces no adverse side-effects. Suggestions have been made for improving indoor environmental lighting so as to make such treatment unnecessary. Two problems stand in the way at present. First, increasing the intensity of indoor lighting by a factor of 10, approximately what would be required, would involve high costs, both in financial terms and in terms of energy consumption. Secondly, such costs would hardly be warranted for a condition which may affect only a small minority of the population.

However, other research has shown that exposing eyes to bright artificial light is also effective in dealing with several additional conditions: sleep disorders, agitated behavior in elderly institutionalized patients, shift work problems, and jet lag. As further evidence accumulates, it seems clear that far more attention must be paid in the future to the health consequences of living and working in environments where light intensity is very much less than in the outdoor environment where the human species evolved.

Susan M. Berta
Department of Geography and Geology
Indiana State University
Terre Haute, Indiana 47809

Timothy A. Gress
ITD/Space Remote Sensing Center
Building 1103, Suite 118
Stennis Space Center, MS 39529

SCS WETLAND RECERTIFICATION USING SPOT DATA

The alarming loss of wetland habitats due to agricultural encroachment has resulted in government legislation mandating the monitoring of wetlands as set forth in the Food and Security Act (FSA) of 1985 (Public Law 99-198, Title 12). Natural wetlands and areas which experience 50 percent frequency of flooding over 15 consecutive days during the growing season can not be converted to cropland after December 23, 1985 without loss of support according to FSA provisions. The 1990 Farm Bill calls for the U.S. Department of Agriculture - Soil Conservation Service (SCS) to conduct periodic reviews and updates of existing wetland inventories. Every year National Headquarters selects 300 counties for State Offices to do a quality review and to recertify a statistical sampling of enough wetland determinations to ensure 95 percent reliability in existing wetland inventories. These reviews identify suspect wetland delineations and alert National Headquarters to potential conversions of wetlands to agricultural lands.

This research proposes a consistent procedure by which multitemporal digital SPOT data may be utilized as baseline inputs for these reviews to assess the changes in spatial patterns of woody vegetation and standing water in northwestern Delaware County, Indiana. High dimensional clustering (200 cluster classes) was used in an unsupervised classification approach as input into a maximum likelihood classification algorithm. The land cover types include open water, wet land/wet soil, and forest/woody vegetation. The final four classified scenes (dates ranging from 1986 through 1989) were incorporated as GIS files and were overlaid and processed to develop additional classified data sets indicating multitemporal changes involving potential forest-to-agricultural land and wetland-to-agricultural land cover changes. The GIS procedures allowed the identification of locations which were: (1) consistently identified as open water (SCS artificial wetland), (2) potentially drained by June, 1989 (non-wetland areas), and (3) potential wetland sites which were overlooked in the previous wetland inventory.

The final assessment in this investigation involved comparing the existing SCS Wetland Inventory digital map with the ISURSL forest and water classifications. The ISURSL base map consisted of forest/woody vegetation and a three date composite of water features where open water and/or wet soil classes appearing two or more times during the time interval from 1986-1988 were included. An overall accuracy of 77.9 percent was identified using the ISURSL classification.

Utilizing digitally classified SPOT data to consistently identify forest/woody vegetation and changes in this land cover proved to be effective establishing SPOT data as a viable data source for consideration in future SCS wetland inventory reviews.

Daryl Blanchett
GIS Manager
Compusearch Micromarketing Data & Systems
Toronto, Ontario

GETTING A GIS TO WORK IN THE BUSINESS COMMUNITY
A VIEW FROM THE INSIDE

The GIS industry has entered an interesting period. GIS has been in existence for some 20 to 25 years and is now widely used in a number of different fields including resource management, academia, municipal governments, and environmental assessment. What is interesting is that after all this time and such a large number of users there is now a huge potential market for GIS technology that may be larger than the existing user base. This new market is made up of all segments of the private business sector including financial, retail, packaged goods, and real estate, and quite possibly the general public as well. This scenario presents a wide range of opportunities to exploit and problems to solve.

The purpose of this paper is to take an objective look at what is in essence an infant industry existing in an already mature one. The paper will examine some of the problems businesses are encountering with GIS technology and discuss how they could be avoided or reduced by learning from the experience of others. Throughout this paper it will be assumed that the need for a GIS to analyze problems and provide answers to pertinent questions has already been established.

It is well known that GIS is an acronym for Geographic Information Systems. What is less well understood or sometimes forgotten is the meaning of this phrase. This term originally referred to an integrated system that used a number of elements all of which had to work together in order for the system to function properly. It is certainly not just the software. For the purposes of this paper a GIS will contain five essential parts. These elements are:

Data: This is the information that will be used in the GIS to answer a set of questions. This is the most essential element in the system. The quality and quantity of data in a GIS are the most important factors in determining the effectiveness of the system. The problem faced by the business community is quite different from the time when GIS were first developed, a wealth of data. Although some people may not be able to envisage a situation where there is too much data, this is certainly true for many cases in business applications. For example, Statistics Canada has available for the 1991 census over 800 variables a 9 levels of geography. In addition most companies have huge databases inhouse that can also be used in a geographic manner. A major part of setting up a GIS is no longer how to assemble or create data but is the selection of which existing data is most relevant for which applications.

People: A GIS cannot operate without trained, qualified people that are familiar with all aspects of the system as well as the purpose of the system and questions being asked of it. The business community is faced with the problem of finding people to run a GIS. In general there are too few people with GIS skills available, but there are even fewer people who have these skills and can also function effectively within the business community. Organizations in the business community that are interested in GIS technology are staffed with people that have training in either business programs or general arts. These people often do not have experience with computers beyond word processing and may not have a good background in geography or analysis.

Applications: For the purpose of this paper the term applications refers to the manner in which the GIS software is used within the integrated system. This could range from developing routines in an advanced macro or programming language to determining the order in which the options or functions of a package will be used to answer a set of questions. Most GIS software packages offer a very good interface with the products functionality. A great deal of work must be done in order to get useful answers from the system. Presently very few software packages provide good tools that the business community can use effectively; but progress is now being made.

Hardware: This refers to the computers that are used to store data, run a GIS and produce output. This is perhaps the are that has gone through the most change since GIS was first conceived but its impact is often ignored. A system that 10 years ago would fill an entire climate controlled room can now sit on a desktop. The new computer systems are now much more powerful, faster, cheaper, and easier to use. There are many more options available for running a GIS. Unfortunately the market is changing so fast that the only thing you can say for certain about buying computer hardware is that if you wait for a little while you can always find a better system at a lower cost. This can make the selection process extremely complex.

Software: This component of a GIS includes all of the computer code that is used to display and analyze the geographic database. It was intentionally left to last because it is arguably the least important component of a GIS. It is still often the part that gets the most attention. There are literally dozens of companies that sell GIS software designed to perform a wide range of tasks. Although it is important to select the proper software package for most simple business applications there are usually a number of packages that are equally capable of doing the job.

There are a number of important issues that the business community needs to address if it is going to enjoy the benefits of using GIS. The first and foremost is that a GIS cannot be bought off the shelf. It must be built! A GIS is a system. It is a construction of a set of parts that must work together in order to be useful. This lesson has been learned over the years by the groups that have been involved with GIS since its inception. The second issue is concerned with the difference in perspective between the business community and the existing GIS community. There is a large gap between what business believes GIS is and what GIS can do. Finally there is a need to provide trained people who have a set of skills that include a background in GIS technology as well as an understanding of what the business community requirements are.

David P. Brennan
University of St. Thomas
St. Paul Minnesota

CONSUMER PERCEPTION OF SMALL TOWN RETAILERS, DISCOUNTERS AND RETAIL COMMUNITIES

Minnesota's small towns have become a battlefield for the three national discounters. Wal*Mart and Target are adding stores while Kmart is adding stores and expanding and remodeling existing stores. Twenty-six, non metropolitan, small towns have seen the number of discount stores increase 41 percent and discount store space increase 79 percent from 1987 to 1992.

The objective of the study was to determine the impact discounters have had on consumer attitudes and behavior and what it portends for the future of retailers and small towns. Three hundred consumers, 60 in the trade ares of five northern Minnesota towns, were surveyed as to their attitudes and behavior toward retailers, discounters and retail communities.

It was found that consumers prefer to shop the closest "major" town. The strength of a town's draw is related to the size of its retail base, proximity to competing larger cities and highway access. Almost two-thirds of those surveyed shop the major town because they either live in the town or are close to town. While most consider their town competitive with similar sized towns, the majority consider it less competitive with larger cities.

Consumers shop at discounters somewhat more frequently than specialty stores. Consumers shop discounters primarily because of low prices (60 percent) and the large variety of merchandise (20 percent). Consumers shopped specialty stores primarily to buy specialty items (52 percent). Other important considerations were having the item the customer wanted, better quality and supporting local merchants.

I. Cimiolino and F. C. Innes
University of Windsor
Windsor, Ontario

A HOLISTIC APPROACH: FROM HEALTH NEEDS SURVEY TO COMMUNITY DEVELOPMENT: A WINDSOR CASE STUDY

Previously, a report was presented on two iterations of a needs survey carried out in the catchment area of the Sandwich Community Health Care Centre, Windsor, Ontario, Canada. The Centre now in operation for a full five years, continues to espouse a philosophy of being consumer driven in its mission. As such, close attention is paid to expressed needs in the client population. These inevitably convey a call for much more than even the holistic approach to health earlier adopted; to include a community development plan. This will include upgrading of the material environment as already attempted in Sandwich Olde Towne, pollution monitoring and control, continuation of Partners in Planting and of the Sandwich Police Patrol.

Given a health providers environment where health promotion and prevention strategies are emphasized, rather than curative intervention, the task becomes insurmountable. Almost all areas of life and jurisdictions become involved, suggesting the need for defining the boundaries very specifically. If this is not done topically, perhaps a discrete geographic area can be employed with some hope of success. To this end empowerment is being encouraged to facilitate the building of consensus on needed actions, and staged goals are being established to highlight progress.

J. S. Cirillo
J. C. Penney Company, Inc.
Piano, Texas

STRATEGIC CONSIDERATION TO STORE LOCATION ANALYSIS IN MEXICO

J. C. Penney, a national department store with over 1,200 stores in the United States and Puerto Rico, is proceeding with plans to open stores in Mexico. As part of this expansion, a store positioning strategy has been developed to assist the Company in choosing the most viable locations for stores.

The strategic approach initially focuses on targeting the largest metropolitan areas in the interior of the country, so as to avoid sales diversion from U. S. border stores. The Company, in evaluating the various storing alternatives being offered, considers the most affluent sectors within a market for specific store locations.

One overriding challenge to conducting strategic area research in Mexico is the lack of reliable demographic data, particularly on a small area level. Other differences include: the convention when classifying income to use multiples of the monthly minimum wage; the size and importance of the informal or underground economy; and differences in consumer purchasing behavior. How J. C. Penney and the Real Estate Strategy Development and Research Department are addressing these concerns will be the focus of this presentation.

Mark W. Corson
Department of Geography and Environmental Engineering
United States Military Academy
West Point, New York

STRATEGIC SEALIFT IN THE GULF WAR

The purpose of this paper is to analyze the strategic sealift requirements of Operation Desert Shield from a geographic perspective. My intention is to identify lessons learned and analyze their potential impact on future military contingency operations. Strategic sealift is fundamentally geographic in nature. American military forces were for the most part located at bases in the continental United States (the supply location and transportation origin point). These forces were needed in Saudi Arabia (the demand location and transportation destination point). The intervening distance between origin and destination points was 8600 miles. Speed of movement was critical. Fortunately, the excellent Saudi ports and airfields were intact and available to the American forces since these facilities were essential for the quick and efficient entry of U.S. forces into the area.

Military land forces consist of many types of units that broadly fall into the categories of "heavy" and "light" forces. Although light forces enjoy an advantage in restricted terrain such as jungles, they are at a severe disadvantage in open terrain such as desert. Heavy forces consist of tanks, and mechanized infantry units that have great mobility, firepower, and armor protection. It was essential that these heavy forces reach Saudi Arabia quickly to provide a credible deterrent to further Iraqi aggression and subsequently an effective offensive force capable of liberating Kuwait. Heavy forces have far too much equipment to move by air so most of it must be moved by ship.

The supply of military units was located predominantly in the continental United States. Some materials were located on ships that were pre-positioned at Diego Garcia in the Indian Ocean and Guam in the Pacific to support this type of operation. Additional forces were forward based in Europe. All of these units were needed in Saudi Arabia. Given the geographic problem, there are several possible solutions including: chartering U.S. Merchant Marine vessels, chartering foreign flag merchant vessels, using ships of the Ready Reserve Fleet, and using Fast Sealift Ships. Other solutions include afloat pre-positioning and forward basing.

This paper examines each of these options in light of the lessons learned from Desert Shield. Sealift options take the longest time, but also provide the most flexibility in terms of reacting to crises in different locations. Many problems must be overcome, however, such as ship availability, crew availability, and the limitations of foreign flag vessels. Pre-positioning on ships and forward basing are the best solutions to overcome distance, but their effectiveness is limited by where they are placed and how much they can carry.

The findings of this study suggest that in a changing world with multiple, widely dispersed potential threats, flexibility is desirable. Desert Shield was a major test of our sealift capability that demonstrated several critical facts. The U.S. cannot conduct a major overseas military deployment unilaterally. We would need foreign flag vessels and foreign crews. Continued degradation of U.S. strategic sealift capability and reliance on foreign ships and crews adversely affect the independence of U.S. foreign policy by limiting our ability to project force. The best army in the world is useless if it cannot get to the battlefield. Our forces will be limited to acting on this continent if they do not have the resources to overcome the tyranny of distance.

Doug Cuthbert
Environment Canada, Burlington

"AN INTERNATIONAL INVESTIGATION OF GREAT LAKES WATER LEVELS"

This presentation summarizes an investigation into fluctuating Great Lakes water levels over six years, 1987 to 1993, by an international team of up to 140 technical experts and lay people - many of whom were geographers. Record high levels occurred on several of the Great Lakes in 1985-86 and a severe storm in December 1985 caused considerable shoreline flooding and erosion damage. As a result, the Canadian and U.S. governments asked the International Joint Commission to lead an investigation into ways of "mitigating the adverse consequences o fluctuating water levels on the Great Lakes."

The water levels of the Great Lakes fluctuate according to annual cycles and year-to-year trends in regional water supply. Annual and 20th century period of record water level ranges of 0.4 and 1.0 meters respectively are observed on Lake Superior. These water level fluctuation ranges increase as one proceeds downstream with Lake Ontario having annual and long term water level ranges of about 0.5 and 2.0 meters respectively. Only two of the Great Lakes, Superior and Ontario, have control dams on their outlet rivers which have some moderating effect on level fluctuations for these two lakes. The other Great Lakes, Michigan, Huron and Erie, have no structural controls on their lake outlets.

The driving force behind this investigation was a coalition of shoreline property owners from Canadian and U.S. shorelines of the lakes, particularly Lakes Erie and Michigan. These riparians had experienced considerable property damage and discomfort during the record high levels of 1985-86 and firmly believe that the construction of outlet control dams to regulate the levels of the Great Lakes will resolve the fluctuating lake level and shoreline damage problems.

Opposed to further structural controls and manipulation of lake levels are the "environmental" interests who contend that naturally varying lake levels benefit the Great Lakes ecosystem and natural wildlife, fisheries, wetlands and biological communities. Either sharing or between these opposing viewpoints are eight other interest groups, categorized as representing agriculture, commercial fisheries, commercial navigation, industry, municipalities, native peoples, recreation and tourism, and hydropower.

The major components of this lake level investigation focused on public participation and information, land use and management, lake level regulation methodologies, crisis condition responses, guiding principles, measures evaluation and implementation considerations. Of particular interest to the geography community are components of the study devoted to identification and quantification of the areal extent of flooding and erosion around the Great Lakes, assessments of damages attributed to flooding and erosion, frequencies of high and low water levels combined with wave and storm setup of the lakes, the potential impacts of climate change on the watershed, crustal movement/uplift, and the development of land use and management practices that might serve to minimize future shoreline damages associated with fluctuating lake levels. Of considerable interest and public debate was the relationship between water levels and shoreline erosion.

Study conclusions and recommendations as contained in a public newsletter and technical report released in April and June 1993 have been presented to the International Joint Commission. These recommendations encompass a number of areas of interest and work for the geography community of the Great Lakes including erosion monitoring, potential damage surveys of shorelines, hazard land identification and acquisition, floodproofing, erosion setbacks and a land use and shoreline management program.

Following a 90-day public comment period through September 1993, the Commission is formulating their own recommendations and report on the issue to the Canadian and U.S. governments.

April Davis
GeoDemographics, Ltd.
69 Arch Street
Johnson City, New York 13790

**APPLICATION OF S.D.S.S.
MODELLING IN UTILITY MARKETING**

Geographic Information Systems (GIS) have come a long way since punch cards. However, the thought-to-be endless path of GIS has been cut off by the evolving Spatial Decision Support System (SDSS). Not only does the SDSS model incorporate the classical components of the GIS (retrieval, input, analysis, mapping, and output), but it also addresses the goals of the end user. SDSS models are currently implemented in strategic end markets such as the utility industry.

This paper discusses the SDSS model and the targeted application of an SDSS in utility marketing.

Paul J. Densham
National Center for Geographic Information and Analysis
and
Department of Geography
State University of New York at Buffalo
Buffalo, NY 14261-0023

A SPATIAL DECISION SUPPORT SYSTEM FOR OPTIMIZING RETAIL BANK DELIVERY NETWORKS

This paper discusses the development, implementation, and use of a spatial decision support system (SDSS) for optimizing retail bank delivery networks. A key element of this SDSS is a geographic information system (GIS) which is used to manage and display georeferenced data. A suite of analytical capabilities, including spreadsheets and locational analysis software, supplement and complement those in the GIS to yield a loosely-coupled SDSS.

While the SDSS has been applied to a range of delivery network optimization problems, this paper focuses on the use of locational analysis to generate alternative delivery network structures. Several types of data are used as inputs to these analyses, including socio-economic data, financial analyses of various branch designs and product mixes, and consumer preferences collected by survey. The types of analyses supported by the SDSS and their linkage to the architectural design of individual branches within a delivery network are discussed.

Richard A. Earl
Roberto B. Anaya
Southwest Texas State University
San Marcos, Texas 78666

MANAGING AND MODELLING A DILEMMA: SPRINGFLOW FROM SAN MARCOS SPRINGS AND THE ENDANGERED SPECIES ACT

The spring fed San Marcos River is a major scenic resource in central Texas and habitat to four federally designated endangered species. San Marcos Springs are the lowest natural surface outlet for the San Antonio section of the Edward's Aquifer. The springs have never dried up, even during the 1950s drought when discharge decreased to a minimum of 46 cubic feet per second. During the 1950s, pumpage from this section of the aquifer averaged 237,000 acre feet (AF)/year and has increased to 478,000 AF/year. Under present pumpage rates it would seem likely that the springs would dry up under drought conditions as severe as those of the 1950s. In 1991 the Sierra Club brought suit in federal district court against the Guadalupe Blanco River Authority, the state agency responsible for managing the surface flow of the San Marcos River, to force limits on pumping in order to protect springflow and habitats of the endangered species. The court ruled in favor of the Sierra Club and ordered the state of Texas to develop an aquifer management plan by June 1, 1993 that would limit pumpage from the aquifer. Effective September 1, 1993 the Edwards Underground River Authority will implement a plan to limit pumping to 450,000 AF/ year and to 400,000 AF/year after 2007 AD. Even before this compromise management program was enacted, persons concerned about San Marcos Springs sought to develop a model to predict springflow from readily available data.

The authors progressed through a series of linear multiple regressions to identify the most useful springflow predictors from the eight county region. Among the variables evaluated were: county precipitation, recharge by surface water sub-basin, pumpage by surface water sub-basin, total discharge by county (springs+pumpage), springflow from selected other springs in the basin, and recharge -pumpage. As shown in Table 1, the recharge from the Blanco River is the best single predictor of springflow. The data verify the significance of the Knippa Gap between Bracketville/Nueces River-west and Uvalde in isolating the westernmost portion of the aquifer. In addition to the Blanco River, the region immediately west of San Antonio appears to contribute the most groundwater to San Marcos Springs. Surprisingly, we found weak to no relationship of pumpage to springflow, which seems contrary to intuition. This disparity could possibly be elucidated by analyzing the relationship of piezometric surface to springflow and correlating total basin pumpage and Blanco River recharge. This type of relationship is hinted at by data from the Medina River sub-basin. Medina River sub-basin recharge has a correlation coefficient of $r=0.74$ whereas recharge minus pumpage has a $r=0.79$; on the Blanco River recharge has a $r=0.85$ whereas recharge minus pumpage has a $r=0.84$ which demonstrates a lack of relationship of nearby pumping to springflow due to the high permeability of the aquifer. Presently we are analyzing the relationship between piezometric elevation throughout the basin, analyzing shorter term relationships, and analyzing the effects of different lag times between recharge and other variables with San Marcos springflow.

Table 1. Summary of multiple linear regression of aquifer variables to San Marcos springflow.

Variable	highest correlation/ location	lowest correlation/ location
precipitation	0.63 San Marcos	0.34 Bracketville
recharge	0.85 Blanco River	0.32 Nueces River-west
pumpage	<0.20	-----
total discharge	0.84 Blanco River	0.34 Nueces River-west
recharge-pumpage	0.84 Blanco River	0.32 Nueces River-west
2 variables	0.93 Blanco recharge+ Comal River total discharge	-----
3 variables	0.94 Blanco River recharge+ Comal River total discharge+ Bexar County total discharge	-----

Glenn Featherstone
DYLEX LTD.
Toronto, Ontario

THE ROLE OF BUSINESS STRATEGIES IN THE DEVELOPMENT AND IMPLEMENTATION OF RETAIL LOCATIONS MODELS

While the interplay of supply and demand work to identify the highest investment use for a site, it is the retailer's understanding of their positioning/business plan which must dictate the type of locations on which to focus a location strategy. An understanding of the principle business and marketing objectives of a particular venture will guide the fundamental development of a location strategy. Where the goals and objectives are misunderstood or badly implemented, locations which would not support a business may be overlooked. Worse still, locations which cannot compete at the highest level of productivity may be chosen. Given the high cost associated with correcting locational mistakes, those who make more than their fare share of poor choices are likely to lose in the marketplace to those who have more accurately and effectively chosen and executed the correct location strategy. In this paper, the fundamental variables required in determining the correct location criteria are identified and examined. The location implications for two alternative types of business requirements are developed.

Lee Feliciano
Pizza Hut Canada

SELECTING THE APPROPRIATE HARDWARE AND SOFTWARE TECHNOLOGY FOR YOUR ORGANIZATION'S DISTRIBUTION NETWORK PLANNING NEEDS

The task of selecting the appropriate technology, be it hardware or software, to address the retail planning needs of an organization can often play a crucial role in the organization's future attitude towards additional technology investment. A balance must be struck between investing in "dead-end" technology, which may be too simplistic or inflexible for an organization to grow with, and investing in "leading edge" technology which is often expensive and can be rendered obsolete within a relatively short time period by cheaper, more effective products that follow. Although not intended to serve as a comprehensive review of available retail development planning tools i.e., GIS, canned databases, etc. the presentation uses Pizza Hut Canada as a case study to illustrate how a medium-to-large Canadian firm struggles with the question of "How much investment in technology is appropriate for our needs?"

Brady Foust
Department of Geography
University of Wisconsin-Eau Claire
and
Howard Botts
Department of Geography
University of Wisconsin-Whitewater

STREET/ROAD INTERSECTION DENSITY AND POPULATION: UNINTENDED USES OF THE 1990 TIGER FILES

Street intersection density was first used by Borchert (1961) provide an objective measurement of the rural/urban fringe. Although theoretically and methodologically attractive, this approach had serious practical limitations: 1) the manual counting of intersections involves time and is subject to error; and 2) available topographic base maps necessary to cover a given area vary considerably in age. Fortunately, the electronic street network afforded by the 1990 Bureau of the Census Tiger Files makes it possible to replicate Borchert's approach easily and quickly.

We used street/road intersection centroids derived directly from the TIGER CD-ROM. The intersection density (per square mile) was correlated with 1990 population at the census tract level for the entire state of Wisconsin. This provides statistically valid measure of the relationship between population and intersection density. Because development (and subsequent population growth) follows the development of the road network (and not vice versa), the calibration provides a quick and easy method of projecting population change resulting from expansion of the road network at the fringes of existing urban places. Such a method has a variety of applications for both the public and private sectors. It provides local planning agencies with a tool to better predict the location and timing for new services such as utilities, schools, and police/fire facilities over the coming decade. Private businesses can use it to more accurately predict housing starts and new retail site opportunities.

Carmine P. Francella
The Bank of Nova Scotia

THE IMPORTANCE OF DESKTOP MAPPING WHEN REVIEWING A CLOSURE OPPORTUNITY

The location analyst is often faced with the dilemma of whether or not to close a retail outlet. Traditionally, unprofitable outlets were earmarked for closure; however, increased competition and lower growth rates are forcing retail chains to consider widespread network downsizing. The location analyst must now be able to select those outlets that, upon closure, would have the least impact on business volumes. The emergence of desktop mapping as a powerful site selection tool has provided valuable insight into the closure problem. Desktop mapping allows the location analyst to take full advantage of customer databases which were previously under-utilized. This presentation focuses on a branch that was a closure candidate and the analysis that followed.

Luisa M. Freeman
Vice President
Applied Energy Group, Inc.
4350 N. Fairfax Drive, Suite 480
Arilington, VA 22203

ELECTRIC UTILITY ADVANCES IN THE USE OF GEOGRAPHIC TOOLS FOR MARKETING APPLICATIONS

Electric Utilities are under tremendous pressure today to cut costs, optimize the performance and anticipate ways to meet customers needs. They are in a critical transitional phase from a period of relative stability with limited growth, few new power plants, and intense competition, seeking out of optimal supply side solutions, and development and offering of new products and services to an increasingly demanding customer base. Some utilities are seeking help toward finding a solution to those pressures in geographic tools. This paper discusses Spatial Decision Support Systems (SDSS) for marketing and customer service applications. This paper will outline the major issues being faced by utilities and how SDSS can help solve them. Some specific utility examples of the use of this system will be described, along with the emerging role of industry organizations.

Milford B. Green
and
Stephen P. Meyer
Department of Geography
The University of Western Ontario
London, Ontario, Canada, N6A 5C2

CANADIAN INSTITUTIONAL INVESTMENT AND REGIONAL DEVELOPMENT

An institutional investor is defined as an organization engaged in investing in securities. Canadian institutions (such as banks, insurance companies, financial institutions, public and private pension funds, corporations and charitable foundations) have become a dominant influence within the nation's economy for primarily three reasons.

(1) The capital accumulations of, particularly North American, institutions investing in Canada has become large.

(2) The block-ownership structures, common to most Canadian corporations, give institutions the ability to control more easily the operations of their investments in comparison to a situation where the shares are more widely-held (such as in the United States).

(3) Not only is there the ability, but there now exists the inclination by the institutional investor to become more involved and, when appropriate, over-rule corporate managers in this age of *shareholders' activism*.

Governments in Canada have understood the influence of institutional investment and, particularly with publicly-owned pension funds, have recognized the potential of using this massive wealth in a regional/economic development capacity. The overall emphasis of this paper is to contrast two such applications: the Quebec government's utilization of the *Caisse de Depot's* pension-funded financial reservoir to maintain Montreal's industrial base and, through the *Ontario Investment Fund*, the proposed use of pension funds in Ontario to stimulate province-wide economic development. Interestingly, both the Quebec and Ontario scenarios closely conform to a growth-pole/centre regional development strategy.

It has been widely recognized that Toronto and Montreal are the growth-centres of their provinces' respective economies. And (as is shown in the empirical section of this paper) like most other forms of economic activity, the vast majority of institutional investment activity both originates and settles within Toronto and Montreal in comparison to other Canadian cities. Thus, the Quebec government's policy to use the *Caisse* to promote Montreal's economic well-being is as much as effort to keep the entire province's economy stabilized. this emphasis on Montreal and the fact that the *Caisse's* most significant investments are made within large, financially stable companies are two likely explanations for the institution's success in a provincial development role.

In Ontario, with the *Ontario Investment Fund*, Toronto is depicted as a growth-centre with perhaps too much influence and one that has excessively polarized industrial activity. As a result, with the fundamentals of growth-pole/centre theory still intact, a policy distinctly opposite to the Quebec experience is being proposed. The New Democratic Party (NDP) controlled provincial government is targeting institutional funds towards higher value-added activities that are not necessarily located in the greater Toronto area. Provincial development will therefore not come through Toronto's *trickle-down* effects, but by direct peripheral subsidization.

In summary, the focal points of this paper are as follows. First, to more completely discuss the affluence and influence of institutional investment in Canada. Second, by noting the locations of both investor and investee companies, to illustrate Canadian institutional investment activity spatially. Third, to confirm the link between institutional investment and Canadian regional/economic development policy with specific reference to the *Caisse de Depot* and *Ontario Investment Fund* applications.

William J. Gribb
Department of Geography and Recreation
University of Wyoming

Wind River Reservation Land Consolidation for Management and Development

In 1984 Congress passed a bill allowing the Wind River Reservation to determine and separate lands between the Arapaho and Shoshone Tribal groups. The 2.8 million acres has a wide range of resources and an uneven distribution of settlements. Six major criteria were established to examine several land consolidation scenarios. The six major criteria are: historic tribal lands;

contemporary tribal lands; natural resources; agricultural potential; accessibility; and contiguity. Six different scenarios of consolidation have been developed varying the weightings of each criteria. The scenarios are based on different assumptions which weight on a spectrum from cultural preference to economic preference. The scenarios vary from presenting larger amounts of land in the west for economic development while the east receives more land for cultural development. However, non-resource development has a higher potential in the eastern portions of the Reservation because of the accessibility to transportation. The combination of land ownership differences between two tribal groups along with the interspersed Anglo ownership makes this a difficult project to solve. This is further complicated by the uneven distribution of natural resources and agricultural potential.

Rhonda Groothousen
Department of Geography
University of Wisconsin-Eau Claire
Eau Claire, WI

GIS AND REGIONAL CLASSIFICATION

Regionalization plays a very important role in geography. A precise definition of regions is often required for policy decisions because different regions require different approaches. Typical problems associated with regionalization include: 1) data division and grouping decisions; 2) the distinction between absolute and relative variables; and 3) variation in areal unit size. Two quite different dairy regions exist in the United States and the differences between them are accelerating. They are often treated the same by policy makers, with the traditional family farm region of the Upper Midwest being subject to the same policies by the USDA and other agencies as the larger corporate industrial dairy operations of California.

A Geographic Information System (GIS) was used in this study to highlight quantitative differences between the two regions. A GIS can be used to quickly and easily solve the problems of data division, raw vs. relative variables, and variations in areal size. Three variables were used to define the two dairy region types in the United States. They were: 1) the number of farms with milk cows; 2) the number of milk cows; and 3) value of dairy products sold. Regional distinctions were then analyzed through maps and tabular data to reveal the differences between dairy region types.

Vincent P. Gutowski
Department of Geology/Geography
Eastern Illinois University
Charleston, IL 61920

BENEFITS AND PROBLEMS IN USING A SIDE-CHANNEL RESERVOIR CONFIGURATION CHARLESTON, IL

In seeking to improve the municipal water supply, the City of Charleston decided on a side-channel reservoir configuration to replace the traditional dam/spillway river impoundment it had used for over 30 years. The result was a water supply with an increased storage capacity and a longer expected life span. Although fluvial sedimentation rates declined, shoreline erosion rates increased. Without the throughflow capabilities of the old river impoundment configuration, the new reservoir serves as a sink for nutrients and organics, promoting nuisance algae blooms and reduced dissolved oxygen levels within the reservoir. Low-cost, low-impact, non-constructive methods are being tested to determine if sediment and nutrient inputs can be reduced.

R. Craig Ham
Department of Geography and Environmental Engineering
United States Military Academy
West Point, New York

U.S. SECURITY INTERESTS IN SOUTHEAST ASIA

In the aftermath of the bipolarity of the Cold War, global security policy of the United States is shifting from that of containment toward a regionally focused policy that calls for constructive engagement with a variety of partners. When last articulated, our vital interests were to promote the following four objectives:

- The survival of the United States as a free and independent nation with its fundamental values intact and its institutions and people secure;

- A healthy and growing U.S. economy to ensure opportunity for individual prosperity and resources for national endeavors at home and abroad;
- Healthy, cooperative and politically vigorous relations with allies and friendly nations;
- And, a stable and secure world, where political and economic freedom, human rights and democratic institutions flourish.

These interests are vital to our continued status as an economically powerful country that depends on freedom of the seas for the unencumbered exchange of trade; commercial access to a region that is emerging from a "developing" status to take its rightful place in the economic hierarchy; and the prevention of the emergence of any hegemonic power or coalition that could threaten the fledgling democracies and emerging economies.

My recent travels reaffirmed that U.S. policy toward Asia in general, and Southeast Asia in particular, is of paramount importance to our continued national well-being. Reason tells us that the United States, by virtue of both geography and history, is a Pacific Power with enduring interests in the Asia-Pacific region. With respect to the economic aspect of these vital interests, more than 35% of U.S. international trade is with the Asia-Pacific region, more than any other region of the world; more than with the European Community or with our American neighbors. Economics and security, however, cannot be separated when examining this region.

Discussions with senior military and civilian security officials in several Southeast Asian capitals reinforced the view that the role U.S. military forces can play in this region can be viewed in three contexts. The first is the maintenance of sufficient forces to convince our allies of the sincerity of our intentions with respect to mutual defense agreements, capable of engagement and participation in peacetime operations. The second aspect is that which affords deterrence and cooperation in times of crisis; and the third guarantees an opportunity to gain victory in times of conflict -- times that have almost never been of our own choosing with respect to time and place. The conditions there today are such that we ought to be wary of the likelihood of the emergence of powers or circumstance inimical to our vital interests; and thus we feel compelled to maintain the power base necessary to ensure these lofty aims.

This paper examines the U.S. security response in an effort to preserve these vital interests. Current U.S. security policy in the Asia-Pacific region has its roots in the Viet Nam conflict and its aftermath. From a basic concept of the primacy of Asian responsibility, supported with appropriate U.S. resources, came the 1990 East Asian Strategy Initiative (EASI) which refocused U.S. security efforts in Southeast Asia from a permanent base structure to a series of less permanent, more flexible arrangements providing periodic access to facilities. The world of the twenty-first century is sure to be one of increased interdependence among states. This examination concludes that our own fate with respect to security issues and economic growth and prosperity is inextricably linked to the political progress and economic growth of others in the region; and that our friends and allies there see a strong American security presence as a stabilizing force.

Dorothy Hanson
Good Samaritan Medical Center
Pheonix, Arizona

USING ARIZONA ZIP CODES TO TARGET AREAS OF HIGH INCIDENCE OF NEWBORNS EXPOSED TO DRUGS IN-UTERO

A statewide study to determine the prevalence of newborn drug exposure carried to a live birth was conducted by the Office of Women's and Children's Health of the Arizona Department of Health Services. Drug screening of newborns by meconium (first stool of newborn) analysis was performed to determine the prevalence and epidemiologic characteristics of drug use by women giving birth in urban and rural areas of Arizona. This data then points to those zip codes with the highest incidence for targeting programs on drug use in pregnancy.

The use of drugs harmful to the mother and fetus is a growing problem nationally. The prevalence of the problem in Arizona is assumed significant. However, until this study, there were no statistical data to support this assumption. The National Association for Perinatal Addiction Research and Education reported on a survey done in 1989, that as many as 375,000 children are born annually in the United States with prenatal drug exposure.

Arizona Hospitals that normally deliver a minimum of 150 babies each year were recruited and participating hospital staff received procedural training via telephone and in writing. The laboratory also provided technical assistance to hospital staff during the collection period.

For a specific number of newborns, in the spring of 1993, each hospital collected meconium specimens. Southwest Laboratories analyzed these specimens by an approved immunoassay for the metabolites of Amphetamines, Cocaine, Cannabinoids, Opiates, Phencyclidine, and Barbiturates.

Demographic data were collected on the form provided as follows: mother's age, race, ethnicity, residence area (zip code), payor, newborn gender and birth weight. The identity of the subjects in the study is protected by the omission of names and addresses on the forms. Specimen number and form number were matched for identification purposes.

The results of the study are reported by statistical data relative to the prevalence of drugs used by pregnant women in all geographic areas of Arizona using zip codes. Comparison of each drug-exposed group to the drug-negative group was done by Chi-square analysis, Student's t-test, one-way analysis of variance or another appropriate statistical method. $P < .05$ was taken as the level of statistical significance. The report contains statewide prevalence levels, prevalence by counties, prevalence by urban or rural locations, prevalence by zip code and prevalence by hospital level.

Tom Haythornthwaite
Geography Division
Statistics Canada
Ottawa, Ontario

1996 CENSUS GEOGRAPHY RESEARCH

The Geography Division of Statistics Canada faces a number of challenges as it prepares for the 1996 Census. The most significant of these are:

(1) Increasing the effectiveness and efficiency of our collection mapping program. As part of this program, we must create unambiguous Enumeration Area maps from digital and analogue sources for 46,000 enumerators, many of whom have little or no experience using maps.

(2) Identifying the areas of major residential growth. It is our goal to delineate Enumeration Areas in which enumerator workloads are as similar as possible. To accomplish this, we must know which areas are growing rapidly, so that we do not devise Enumeration Areas that contain more households than a single enumerator can handle effectively.

(3) Increasing the effectiveness of our census geographic products and analytical tools. These include atlases, reference maps and public on-line GIS query systems.

The first two goals directly affect Statistics Canada's internal operations, although the ultimate benefit is a more accurate, cost-effective Census for all Canadians. The last goal, however, reflects the importance Statistics Canada places on assisting data users across Canada in their various endeavors that require census data. To meet all of these objectives, the Geography Division is being assisted by a nationwide community of geographers and geographic data users.

J. Harris and J. G. Nelson
Toronto, Ontario, Canada M5B 2K3

A TOTAL ECONOMY VIEW OF TOURISM AND RECREATION DEVELOPMENT

The meaning of "development" is hidden in a tangle of words. In the name of progress, development programs cancel the identity of communities and impose on them processes of change that create dependency on external markets. Because development is an illusive entity -- difficult to analyze, difficult to implement -- economic concepts and theories have been highly influential in shaping development planning. Development programs have suffered from the abstractness that has prompted criticism of economic theory and practice (Daly and Cobb 1989; Gorz 1989). This paper presents an alternative view of development, based on a model of the local economy which includes formal, informal and vernacular sectors. The understanding that the economy has two or three sectors, beyond that of the market, is not new. Chayonov, Sahlins, and Polanyi are among those who have described a three-sector economy (de Romana 1989).

Total economy profiles of the community or region provide more complete information on the nature of the human ecology as a basis for another "development." We have utilized a total economy profile in our study of the human ecology of a mixed fishing and tourism community in Java (Harris and Nelson 1992) and further research is underway in the Temagami area in northern Ontario (Harris 1993). We have found that these profiles provide a better understanding of the linkages between economy, society and environment and a foundation for evaluating the impacts of development programs and policy. The implications of tourism and recreation development initiatives for activities in the vernacular, informal and formal sectors and the effects on relationships between these sectors must be understood if we are to make informed choices among development alternatives.

Brian Hoefle
INTELLIMED
Phoenix, Arizona 85012
and
Rena J. Gordon
Department of Family & Community Medicine
University of Arizona, Tuscon, Arizona and
Department of Geography
Arizona State University, Tempe, Arizona

USING PUBLIC HOSPITAL DISCHARGE DATA AND GIS TECHNOLOGY FOR MAKING INFORMED HEALTHCARE DECISIONS

There is always a need for accurate data for health planning. In 1983 the Arizona Hospital Discharge data set became available and its potential value led to the formation of INTELLIMED, a privately owned company. Early reports consisted of comparing hospitals on lengths of stay, average charge per case, and case mix index. Products have expanded to include reports on physicians, patent migration, and market share, using a PC based package called the INTELLIMED Answer System. Customized products and services are available to clients, who are predominantly hospitals, health plans, physician groups, and a major employer. The client base is expanding to include rural hospitals, health care suppliers, health care purchasers, and health care consulting practices.

INTELLIMED serves clients in six states and expansion into those states which provide the necessary data sets are underway. Although over half of the states in the country require hospitals to report on discharge activity, a far fewer number collect a physician identifier and release it as part of the public data set. The trend is clearly towards requiring hospitals to submit data, or increasing the set of data to be reported and making that data available to the public.

The PC based INTELLIMED Answer System is actually three separate data base systems, one of which is the geographic information system (GIS). This system stores data in dBase format so that it may be cleaned and sorted geographically and then displayed on a map using the Atlas PRO software. This paper describes the System's key data elements in the following most commonly used data sets: (1) hospital discharge data file; (2) census tract data file; (3) board of medical examiner's physician data file; (4) DRG data file. Samples of the types of reports generated are presented and demonstrate the feasibility and usefulness of this tool.

Mapping is used for example, to identify a hospital's primary service area based on market share. The analysis is further refined by mapping market share for a selected payor group or specialty. Demographic data is mapped to identify concentration of selected population and rates of growth. Health plans use the mapping product to analyze the location of providers to the client base, thereby delivering healthcare more effectively.

Results are discussed in terms of the Answer System's utility in health care decision-making. As INTELLIMED looks to the future, the focus will be on moving into new states as the data becomes available. INTELLIMED is also considering a product offering for those states with a more limited data set. Data sets on utilization review and outpatient activity are under review for inclusion. Staying abreast of the technological developments in the PC arena is a challenge. INTELLIMED plans to continue to identify developments which will enhance the usefulness of geographic based data for health planning.

Bruce Kerr and Richard Kellar
Vice President and Senior Researcher
Retail Group
Compusearch Micromarketing Data and Systems
Toronto, Ontario

GIS BASED RETAIL CANNIBALIZATION MODELS

Compusearch has long been involved in the development of site evaluation models for retail clients. These models have ranged from simple ones based on scoring and analogue methods to very sophisticated spatial interaction models for predicting sales at existing, modified or new stores. There are numbers of other firms which are in the business of constructing high end site selection and evaluation models for clients. However, these high end models are not best sellers. Even large retailers who appreciate quality research are reluctant to spend $50,000 or more on a site model, especially when such models may only be suitable for a particular market.

One of the most important motivations for wanting a site model is to be able to answer questions like: "What will be the impact on sales of existing stores in our chain when we add a new store in a market?" and "What will be the impact on nearby store sales of closing an existing store?" The questions are often phrased in terms of "cannibalization." For questions of sales projection, absolute predictions are required. Retail managements are usually satisfied with relative answers to these impact and cannibalization questions. Having a projection of a 5% loss of sales vs a 15% loss in sales is what is required. It is often the case that relative questions can be addressed with simpler, less expensive approaches than absolute questions. This is the case with site models.

Compusearch has developed a relatively inexpensive modelling approach, which provides good relative answers to cannibalization questions, using data which are generally available or which are appreciated as being necessary. We call the product "Cannibalization Models," and it makes use of GIS, SPANS GIS in particular. The product is being positioned as an important benefit of reconceptualizing trade areas from traditional all or nothing circle and polygon entities to what we are calling "Fuzzy Trade Areas." Fuzzy trade areas are spatial surfaces based on estimated probabilities that people in neighbourhood $\underline{i}$ patronize a store at location $\underline{j}$. A person or neighbourhood is no longer a member of only one trade area but rather is a partial member of several. These gravity type models or Huff type models, of course, have a long history in applied geography. The reason these are not as widely used as one might guess is that they are difficult to use without a lot of expensive custom programming and high level statistical analysis. We use conventional customer spotting data and the "gravity model" module of SPANS GIS to model trade areas, trade area overlap and finally cannibalization impacts for about $1000 per impact analysis study. The product is selling well and expected to be very popular.

In order to develop the model, we obtain customer spotting data from the client on the market being modelled. We have a strong preference for data with the specific stores patronized being identified. These point data are then mapped along with the other point layer - **stores**. The store data should include nearby stores in the chain and also competitors. An experienced retail GIS analyst then selects several distance decay parameters which are expected to be suitable for Huff modelling, and surface maps showing contours of the probabilities of patronizing selected stores are drawn. Experience permits us to select a parameter which generates probability surfaces with the best qualities. Then a different (trade area) surface is generated for each store involved directly in the analysis. (These surfaces need not use the same parameter). The final GIS step is to undertake surface overlay analysis for various combinations of existing and new stores which assigns estimates of demand to different probability classes with respect to the different stores. The pre-impact estimate of demand realized by each store is computed. Then for each added store and each existing store of interest a 10 by 10 matrix of impacts is prepared. The reduction in realized demand comes directly from assigning away some formerly captured demand to the new store. The approach is theoretically rigorous, works very well in practice and is efficiently implemented.

The final report delivered to the client is a set of spotting maps, a set of coloured fuzzy trade area probability maps, a set of three impact tables for each question and a short executive summary.

Tom Kierans, P. Eng.
Grand Canal Project, Newfoundland

WATER RESOURCE ISSUES IN THE GREAT LAKES BASIN

The first paragraph of the International Joint Commission's November 1992 report on the results of its research for the Great Lakes Levels Reference reads "Climate change (the Greenhouse Effect) could reduce water supplies to the Great Lakes-St. Lawrence River basin to the extent that, in the next century, outflow from the five Great Lakes would be diminished as follows: Superior, 13%; Michigan-Huron, 33%; Erie, 40%; and, Ontario, 39%.

No warning can be clearer than this. Even if such a dramatic reduction in Great Lakes outflows is only a possibility and there is full regulation of levels and flows with maximum demand controls and conservation, it would be unjustifiably imprudent to fail to plan NOW for such severe worsening of current basin pollution, wetland and marine life destruction as well as energy, shipping and other problems. U.S. pressure for more water withdrawals like that at Chicago is certain. Montreal Harbor would be unusable. Our only realistic answer is a major, new integrated, interbasin water transfer system to and from the Great lakes.

Unfortunately, misguided policies on both sides of the international border oppose interbasin transfers. Canada's federal policy bans "water export" to many shared basins where most Canadians live. This policy misleads Canadians by suggesting that *all* water transfers reduce our freshwater supplies and cause major environmental harm. This is mainly due to the failure of federal officials to recognize fundamental differences between types of interbasin transfers.

Diversion-type transfers from a donor basin's headwaters (like the proposed NA-WAPA diversion) are obviously unacceptable due to the impacts from flow losses on downstream environments and the flooding of major mountain valleys. Because Canada's federal policy wrongfully claims that *all* interbasin transfers are diversions in rejects *all* transfers. In contrast to diversions, *recycled run-off* transfers, pumped from a sea level enclosure *increase* freshwater supply and enhance *both* donor and receptor basin environments and do not flood mountain valleys. Recycled run-off technology has been used successfully for many years in The Netherlands, California and elsewhere. Such transfers also control floods. Recycled run-off's proven environmental, social and economic benefits,

Canada's regional precipitation variations, and worsening Great Lakes Basin water supply crises make anti-interbasin transfer policies absurd.

James Bay basin's huge run-off is almost twice the flow of the Niagara River. Most of the run-off can be conserved by dikes in this salty and shallow bay. By opening gates in the dikes at low tide and closing them at high tide (as the Zuider Zee was changed to the freshwater IJselmeer), James Bay can be changed to a freshwater lake. Some conserved run-off can be transferred, as needed, via a pump-aqueduct system to the Great Lakes. Concurrent transfers from the Lakes by a new Prairie Transfer Canal to Canada's dry prairies and then to the U.S. are needed for system operation. Over 160 million Canadian-U.S. users of recycled run-off from James Bay would pay $50-$100 per year for this essential service.

Canadian benefits from regulated management of the transfer systems include:

- improved James Bay and Hudson Bay environments by reducing the fresh and salt water stratification that now inhibits biological productivity in both bays:
- an *increase* (about 10%) in Canada's freshwater supplies;
- reduction in both high and low level fluctuations in the Great Lakes basin;
- protection for Montreal and all downstream areas from disastrous flow loss; and
- revenues from a cooperative and profitable Canada-U.S. water supply system.

Phillip T. Kolbe
Department of Finance, Insurance, and Real Estate
Fogelman College of Business and Economics
Memphis State University
Memphis, Tennessee 38152

Janet L. Papers
Sun West Market Research, Inc.
6125 Greenlee Street
Arlington, Tennessee 38002

SMALL TOWN VERSUS BIG CITY: VARIATIONS IN THE PERCEPTIONS OF COMMERCIAL LENDERS

Do loan officers in small towns do business differently than their counterparts in large urban areas? This is the focus of a study which may also give insight into the malaise that has plagued financial institutions. A survey of bank and thrift loan officers was conducted in four major cities, as well as smaller communities in New Mexico and West Texas. The findings reveal that loan officers have varied perceptions and have improperly analyzed some financial information because of misconceptions.

E. Kulke
Geographisches Institut
Universitat Hannover

SPATIAL EFFECTS OF THE ENTERPRISE CONCENTRATION PROCESS IN GERMAN RETAILING

In the past few decades, a major concentration process has taken place in retailing in the Federal Republic of Germany. A small number of businesses are achieving a steadily growing share in sales. The Gini coefficient, representing the concentration of sales shares on individual businesses, rose from 0.63 in 1968 and 0.71 in 1976, to 0.79 in 1990, whereas, between 1968 and 1990, the number of retail enterprises declined by 15.5%. in 1990, 1% of the retail businesses achieved 39.2% of total sales in retailing. In contrast to the 1970s when the number of small and medium sized enterprises decreased, the 1980s saw sales become increasingly concentrated on big enterprises; however, parallel with this development, numerous new enterprises consisting of only one business were being established. These small units supply small local market niches and markets specializing in certain types of merchandise.

The branches of large retail enterprise chiefly offer a standardized assortment of goods which is limited to high demand articles. The standardization of this assortment, in combination with the diversification of consumer demand for open markets offering specific goods, results in limited sales potential, which are of no interest to large enterprises. These markets are supplied by small single-business enterprises which are often run "on the side" and sell highly specialized goods, such as wooden toys, Tiffany lamps, Turkish food, etc. These new, special retailers typically prefer locations on the periphery of urban subcenters.

The branches of the big retail businesses are predominantly large. Their establishment causes the average business size to increase, the number of retail stores to decrease, and the supply network to grow wider-meshed. The branches choose only locations offering a sufficiently great sales potential. In rural areas, they prefer the central places and agglomerations in non-integrated locations.

Small, older business which due to their cost structure cannot compete wit the new branch stores are forced to close. This creates supply gaps in small rural places and in residential areas of large cities and, as a consequence, minor urban subcenters decline in importance. In these areas, local market niches spring up which draw their supplies from new, small types of businesses, such as, e.g., mobile food retailing, kiosks, gas station shops, etc.

Wade D. Laufenberg
J.C. Penney Co.
Plano, Texas

RETAIL MARKET ANALYSIS IN U.S. MEXICAN BORDER MARKETS

The history of retailing along the U.S. Mexican border has been erratic, generally mirroring the economic fortunes of Mexico. Confirmation of the North American Free Trade Agreement would result in significant changes directly impacting retailing, particularly facilities located in border markets. The U.S. border market and its Mexican sister city possess interdependent economies. This relationship is most notably characterized by the extensive development of maquiladora plants which operate under the "in-bond" industrialization program. Correspondingly, the retail sector also exhibits a mutually dependent relationship.

The border market presents unique challenges to retail market analysis. These challenges include securing accurate population and income data for both sectors of these markets and determining the level of Mexican resident sales import into the U.S. market. Retail analysis has been expanded to include assessing the impact resulting from the recently announced plans of a number of U.S. retailers to locate stores in Mexico. The presentation details the methodologies for examining the relationship of the U.S. border market to its Mexican sister city and the subsequent impacts on indigenous retail facilities

Kirk J. G. Lee
Compusearch Micromarketing Data & Systems
Toronto, Ontario, Canada

GEODEMOGRAPHIC SEGMENTATION ANALYSIS USING LIFESTYLE WORKSTATION

Market segmentation has been available for many decades. Segmentation systems categorize the population/households into unique groups or clusters. Segmentation systems can accomplish this by using demographic or behavioral information. As an example, demographic information can be a combination of selected lifestage variables, ethnicity, or any other variables which may be useful in grouping the population into distinct clusters.

Segmentation becomes much more powerful when it is also geographical which in turn guarantees its targetability. Geodemographic segmentation has been used since the mid-seventies and allows companies to find target segments for their product/service by analyzing and identifying areas with the highest potential. This exercise was usually performed at great expense regarding computer time and human resources. However, in the last decade, the advent of higher capacity and more capable personal computers, coupled with major software developments has enabled companies to interrogate and analyze their customer databases easily on their desktops.

One such system is the LIFESTYLES Workstation. LIFESTYLES Workstation enables a company to analyze cluster-coded customer lists or syndicated databases using Compusearch's 70 LIFESTYLES geodemographic segments or clusters. The Workstation will assist a company in analyzing its strategies and applications (direct marketing, site evaluations, etc.) in an effective, efficient and timely manner. In short, the Workstation allows a company to significantly improve its marketing efficiency by providing the analytical tools for evaluating its products and services by:

1) Illustrating the "lifestyles" profile of a company's product/service.
2) Assessing market potential by cluster of the product/service in a selected area.
3) Ranking with other markets.
4) Ranking with other products/services.
5) Correlating with other products/services (syndicated survey information) for cross marketing.
6) Illustrating potentials with thematic maps of both standard geographies and custom areas.
7) Facilitating the creation of own segments by combining the target "LIFESTYLES" from the 70 clusters.
8) Searching for other products/services which appeal to these combinations.

The ease with which the LIFESTYLES Workstation can be used will be shown during a live demonstration.

J.D. Lord
University of North Carolina at Charlotte
Charlotte, NC 28223

POWER SHOPPING CENTERS AND POWER POLITICS: RETAIL DEVELOPMENT IN CHARLOTTE, NC

This presentation reports on three controversial rezoning cases in the northeast area of Charlotte, NC which involved efforts by developers to obtain zoning changes that would have permitted the construction of large, power shopping centers. Power shopping have become a popular type of center for developers, a trend spurred by the financial strength of a number of large store retailers including discount department stores, home center chains, wholesale price clubs and off-price retailers. These centers are dominated by a few large (power) tenants and contain relatively little shop space. Within the shopping center hierarchy, they are positioned between the regional mall and the traditional community shopping center.

Population and retail growth in Charlotte historically have been concentrated in the southern and eastern sectors of the market. However, since the mid-1980s the northeast area has experienced significant population growth with this growth being influenced by infrastructure investment by the city in an effort to relieve growth pressure in the southeast sector of the market. The growth in the northeast sector has generated considerable interest by retailers and developers in the area, but until recently retail developments had consisted of only small neighborhood shopping centers.

The northeast area had undergone extensive land use planning in the 1980s and early 1990s, and in 1991 a comprehensive land use plan was adopted by local government authorities. This plan specified several sites in the area for various types of shopping centers. The commercial focal point of the northeast area was a mixed-use project, University Place, which was begun in the 1980s. The centerpiece of this project was to be a regional shopping center. The site for the center was purchased by the Hahn Company in the late 1980s and efforts were begun to obtain commitments from department store chains to anchor the center. For a variety of reasons, including the weak economy in the early 1990s, problems within the department store industry and the inability to get a strong, local department store chain to commit to the project, the Hahn Company was unable to obtain sufficient commitments from department stores which would have permitted construction of the center. As a result, Hahn proposed an alternative retail development for the site which would have included three large retail stores in a power center format. This plan represented a departure from the original development plan for University Place, and thus rezoning approval was necessary. At about the same time, two other developers filed petitions for rezoning which would have permitted power centers at two additional sites, the Crosland and Cambridge sites, both located within one-half mile of the Hahn site. In addition to these three proposed power center (McAdams-Norman site). As a result, the northeast area was faced with four sites, all seeking to be developed as power centers. The three rezoning requests plus the one site already approved produced the potential for over 2 million sq. ft. of retail space in the market.

A study conducted by the Charlotte-Mecklenburg Planning Commission of the retail space needs of the area concluded that this large amount of retail space was not needed in the market. This planning body then recommended to the Charlotte City Council that only the Hahn rezoning request approved. Recommendations against the other two rezoning petitions (Crosland and Cambridge) were based on the conclusions that the space was not needed in the market and that retail use of the sites was in violation of the land use plan for the area which had been adopted earlier by the City Council. Two public hearings were held on the three rezoning petitions. What ensued was a very controversial battle on the part of the developers of the Crosland and Cambridge sites to gain rezoning approval. The Crosland site proved to be the most controversial of the three proposals, while little opposition surfaced to the Hahn site at University Place. Although it seemed for a while that the Crosland request was in danger of being denied, a last minute vote change by a city council member who had earlier expressed opposition to the rezoning request resulted in the approval of the rezoning petition by a vote of 6 to 4. The rezoning request for the Hahn site was approved without opposition, while rezoning for the Cambridge site was approved by a vote of 6 to 3. City Council members opted to let what they described as "market forces" determine the eventual outcome. It appears that construction will begin soon on the Hahn and Crosland sites as both of these developers have secured commitments from major tenants including discount department store chains and home center retailers. The other two sites are less certain and may not be built as power centers because of the lack of appropriate anchors.

The events which transpired in the northeast Charlotte rezoning cases raise several questions. How much "power" should an adopted land use plan have? (Many people were left wondering about the value of such plans if they can be changed so soon and easily.) To what extent are land use planners able to plan adequately for retail space in a market? (Many planners have little expertise in retail activity and their ability to prescribe the needs of the area could be questioned.) Can retailers and developers do an adequate job at self-regulation in terms of the retail space needs of an area? What are the costs to both the public and private sector of too much retail space? What will be the competitive outcome if two or more power centers are built? It appears that three of the largest home center retailers in the U.S. will soon be competing in the northeast Charlotte market. Either the two or perhaps three largest discount department store chains in the country will also likely be competing in the market.

Louis K. Marion
Department of Geography
Kent State University
Kent, Ohio 44242-0001

SCALE DEPENDENCY AND DATA AGGREGATION FOR SLOPE AND ASPECT MEASUREMENTS ON DIGITAL TOPOGRAPHIC SURFACES

Digital elevation models (DEMs) have become a major data source for computer modeling of terrain surfaces and are of particular importance to a wide spectrum of applications in the geo-sciences. DEMs and their application have significantly impacted the field of digital terrain analysis. Some examples are: 1) Slope and Aspect Mapping, 2) 3D Perspective View Generation, 3) Line of Site Modelling, 4) Analytical Hill Shading, and 5) Watershed Drainage Network Analysis. This data source has expanded the potential for the geo-scientist to further the research areas of his or her field in a substantial way. However, we must recognize that the results of such research are highly dependent upon the scale, resolution, or aggregation of the source DEM.

Many of the computer modeling applications being conducted in the geo-sciences today give little or no consideration to the potential effects that can be introduced into the modelling process when using DEMs of various aggregation levels. This paper examines the scale dependency of digital elevation data among various levels of aggregation with respect to slope and aspect measurements using both the lattice (GRID), and triangulated irregular network (TIN) data structures. Slope and aspect are chosen to be investigated because they represent the most fundamental structural information used in terrain analyses. Understanding how scales impact on slope and aspect measurements, terrain analyses may be conducted to achieve more accurate and reliable results.

Emmanuel K. Mbobi
Department of Geography
Kent State University
Kent, Ohio.

A MULTISPECTRAL EXAMINATION OF THE BUCKEYE BASIN WETLANDS, TOLEDO OHIO

This paper examines three basic sources of remotely sensed images for identification, measurement and classification of wetlands. The three data sources are: satellite images, black and white areal photos, and color infrared photos. These data sources are examined with regard to their accuracies and shortcomings for delineating wetlands. A topographic base map has been used as the basis for accuracy assessment because of the degree of ground truthing involved in making them. All three data sources have been used in analyzing the Toledo Buckeye Basin wetlands in Ohio.
The Toledo Buckeye Basin wetlands are an example of wetlands influenced by human activity. Activities such as road construction, disposal of waste, and a variety of other environmental activities are very common here. With human activities heavily affecting these wetlands, definition and delineation can be very difficult. Difficult because the factors that determine the delineation of wetland parcels are fragmented by human activities, thereby, creating a discontinuity of these features that delineate the individual wetland parcels.

With the results obtained from the study, it is clear that remotely sensed images are excellent for studying wetlands and other spatial features. The number of wetlands obtained through remotely sensed images is similar to those identified on the base map. On the black and white photographs as well as color infrared some irregularities have been detected. The wetland parcels do not correspond to those on the base map. The problem might be attributed to viewing the features only in two dimensions i.e. length by width which creates a considerable difficulty for most objects to be recognized. That could be one of the problems that has made it impossible for more than two wetland parcels to be detected on the black and white photographs and eight wetlands on the color infrared.

This paper examines some of the problems associated with wetland delineation as a result of the definitions associated with them. It also examines the various data sources used in determining the best way possible for the delineation of wetlands. It also discusses the study area and why it is ideal for this study.

Terry Meyer
MarkeTech Systems
Troy, Michigan

OPTIMAL NETWORK LOCATION STRATEGIES USING 'SITE'

This presentation will focus on the challenge of increasing the efficiency of a retail product delivery system, or chain of stores. Inherent in MarkeTech Systems' methodology is the fact that convenience weighs heavily in a consumer's choice of a drug store when he/she is presented with options. Rarely is a given retailer "the only game in town."

The challenge to the retailer is to make itself convenient to consumers while at the same time optimizing the efficiency and profitability of its delivery system. How many stores are too many? How close is too close? Which locations have the greatest net impact to the chain as a whole, not just the store characteristics can a chain improve to increase the likelihood of capturing customers?

MarkeTech Systems' methodology quantifies the complexities of the retail market place, including how consumers choose, and make possible comprehensive market analyses that would be impossible to do manually.

Monica Nyamwange
East Stroudsburg University

TOWARDS ENVIRONMENTAL REHABILITATION IN AFRICA

African countries are facing formidable environmental problems such as desertification, deforestation, soil erosion, uncontrolled urban growth, and energy and food shortages. These problems stem from the interaction of a number of complex factors: colonial intervention, government policies, rapid population growth, unfair land tenure systems, harsh climatic conditions, and an unfavorable international economic situation. The root cause of environmental deterioration is poverty which is locking the African people into a dismal cycle of events. In their efforts to meet mere basic needs of food, shelter and energy, Africans are forced to destroy the very resources on which their survival depends. In many African countries, the rural poor are being driven on to marginal lands where farming cannot be sustained. Still others are driven into already overcrowded slums and shanty towns.

This paper stresses that the destruction of the African resource base can be stopped through the following measures:

1) Safeguarding traditional systems as nomadism and fallow practices which respect the limitation of Africa's sensitive environment and adapt these systems through new techniques to achieve increased production levels needed to support higher levels of population.
2) Giving special attention to the rural majority and providing them with know-how and low cost inputs and economic incentives which will assist them to conserve their resource.
3) Since rapid population growth has led to environmental deterioration in many African countries, it is imperative that measures aimed at reducing fertility levels be pursued. These should include making couples aware of the benefits of having small families which they can adequately sustain, consequences of large families on the standard of living and the environment, encouraging participation in family planning programs and improving the status of women so that they can become full partners in modern sector jobs which are less compatible with child bearing and child raising activities.
4) Alternatively, the sporadic growth of urban cities in Africa towards the last quarter of this century has led to rapid imbalances of ecosystems. Urban centre have spread towards the suburbia and surrounding environments that previously supported large plant and animal life. To curtail such unwarranted imbalances, growth of these urban centers calls for remedies that would preserve existing environmental systems. Secondly, existing urban systems need to focus on environmental preservation techniques by encouraging proper landscaping.
5) Realizing that food shortage in Africa is an ever existing problem, environmental experts need to enhance environmental education to the rural masses. This should include among others: (i) increased research on drought and flood resistant foods, (ii) ways of stimulating the demand for drought and flood resistant foods without necessarily upsetting existing cultures and traditions of individual societies.
6) Some diversification in livestock mix is beneficial. Many pastoralists concentrate on keeping cattle which have high requirements of water and grass. In drought prone areas, environmental degradation results from concentration of cattle along water sources which leads to soil compaction, reduced infiltration and increased soil erosion. Alternatively, the advantages of goats and camels which include lower water requirements, the ability to utilize browse, more efficient forage conversion, and lower susceptibility to some diseases should be recognized.

7) More than 80% of the African population depends on firewood to meet their energy requirements. This has led to the destruction of tree cover which is vital for maintaining soil fertility, keeping back the deserts and for preventing a desiccation of local climates. Two approaches seem appropriate:
   (i) Increasing fuel wood plantations on wastelands and other areas of little agricultural value.
   (ii) Encouraging the use of appropriate and environmentally sound renewable technologies such as solar and biogas power.

Frank H. Quinn

Great Lakes Environmental Research Laboratory
2205 Commonwealth Blvd.
Ann Arbor, Michigan 48105-1593

THE HYDROLOGIC CYCLE OF THE GREAT LAKES

The Great Lakes are one of the world's major water resources containing about 20 percent of the world's fresh surface water supplies. The Great Lakes basin includes an area of about 766,000 square kilometers, of which about 1/3 is lake surface. The basin includes the five Great Lakes, Superior, Michigan, Huron, Erie, and Ontario, and Lake St. Clair and their connecting channels. Lakes Michigan and Huron are considered as one lake hydraulically as they are joined together at the Straits of Mackinac. Lakes Superior and Ontario are regulated. The remainder of the system is naturally well regulated due to the large lake surface areas and constricted outlet channels. This has resulted in the lakes historically fluctuating through a very small range in water levels, about 1.8 m. This presentation examines the relationship between the hydrologic cycle and the interannual and seasonal changes in lake levels.

The Great lakes water levels serve as integrators of hydrologic variations and climatic change. The major process that determines lake level fluctuations is the hydrologic cycle whose components are precipitation on the lake surfaces, runoff into the lakes from their basins, evaporation from the lake surfaces, and the groundwater flows into the lakes. These components, added together, yield the net basin water supply. The component causing the major multi-year variation in lake levels is the precipitation. The runoff, in turn, magnifies the effects of the precipitation. High precipitation leads to high runoff and vice versa. There have been 3 distinct precipitation regimes over the past 130 years; a relatively wet regime ending in the 1880s, a dry regime lasting from the 1880s to about 1940, and a wet regime lasting from 1940 through present. From 1967 to 1987 the regime was extremely wet which led to record high lake levels in 1973 and 1986.

Variations in air temperature, humidity and wind speed also influence lake level fluctuations through lake evaporation and evapotranspiration from the land basin. All other parameters being equal, higher air temperatures tend to reduce water supplies through higher evaporation and evapotranspiration. The annual mean air temperature around the perimeter of the Great Lakes since 1900 indicate three to four distinct temperature regimes: a low temperature regime from 1900-1929, a high regime from 1930-1959, and a low regime from 1960 which appears to have ended in the 1980s.

Michael Ratcliffe
Geography Division
U.S. Bureau of the Census
Washington, D.C.

SMALL URBAN AGGLOMERATIONS

Since 1950, the Bureau of the Census (BOC) has recognized that urban population exists in densely settled clusters outside of census recognized places. This phenomenon, however, is reflected only in the delineation of urbanized areas (UAs). At population levels below 50,000, the BOC's definition of "urban" applies only to the inhabitants of places of 2,500 or more population. Thus, population located adjacent to an urban place outside of a UA may be considered rural even though the settlement pattern is similar to that of the population within the place. The failure to define small urban agglomerations (SUAs) with populations of at least 2,500 and less than 50,000 understates the extent of the urban population in the United States.

The BOC currently is conducting research to determine the feasibility of delineating SUAs. This research initially focuses on the identification of SUAs in selected Maryland counties representing a variety of settlement patterns, levels of metropolitanization and urbanization, economic and demographic characteristics and topographic conditions. The method used in the investigation to define SUAs is similar to that used to define UAs, namely, the qualification of individual census blocks or clusters of blocks on the basis of population density (at least 1,000 persons per square mile) and adjacency to either the central place or other qualifying territory. The SUAs identified fall into three possible combinations of geographic entities:

(1) an urban place together with adjacent urban or rural places and densely settled census blocks outside of any place;

(2) a rural place together with adjacent rural places and densely settled census blocks outside of any place having a total population of at least 2,500;

(3) a cluster of densely settled census blocks outside of any place and having a total population of at least 2,500.

This investigation, however, raises important questions regarding whether high population density in otherwise rural areas necessarily denotes the presence of urban functions and characteristics, in whatever way those may be defined.

J. L. Rice
Holliday, Fenoglio & Co.
Houston, Texas

HOUSTON'S GROCERY-ANCHORED RETAIL CENTERS GEOGRAPHICAL PERSPECTIVES WITHIN THE REAL ESTATE CYCLE

Grocery-anchored retail centers outperform most other types of retail centers in occupancy and rental rates. These centers are the favoured retail property type by most lenders and one of only two types of retail development occurring today throughout the United States. This paper will focus on the findings of annual surveys conducted over the past six years on Houston's supermarket-anchored retail centers, and will analyze the market and locational factors. The discussion will also relate the performance of grocery-anchored retail to Houston's real estate cycles, focusing on the geographical patterns of retail performance during the recovery cycle.

Jennifer Rogalsky
Department of Geography
University of Wisconsin-Eau Claire
Eau Claire, WI

USING GIS TO TRACK AND ANALYZE INDUSTRIAL POLLUTION AND REMEDIATION

Many industries continue to pollute and pay fines for non-compliance rather than taking measures to remedy the problem. Conversely, others exceed environmental standards for reasons such as corporate responsibility, cost effectiveness, fear of more rigorous standards, and public image.

The purpose of this paper is to evaluate how well a simple Geographic Information System (GIS) can be used to acquire, track, and evaluate data, from a variety of public sources, to connect compliance levels of individual corporations with federal, state, and local environmental policies and regulations. A GIS is a very effective instrument to correlate polygon data (policies and laws) with point data (plants and corporate headquarters). In this paper, corporations and plants are ranked along a continuum from good to bad in terms of their compliance record as recorded by various state and federal agencies and by environmental watch groups. These plants are then geocoded by downloading their ZIP codes from an on-line data vendor. Point-in-polygon analysis is then used to determine the spatial correlation between environmental regulations and plant compliance.

Kurt Schroeder
Department of Geography
University of Minnesota Duluth
Duluth, Minnesota

STRATEGY IN THE BOSNIAN WAR

The withdrawal of Soviet power from Eastern Europe, the death of Marshall Tito, and the collapse of the collegial Yugoslav state presidency has led to the current ethnic conflict in the former Yugoslavia. This conflict reflects the armed ethnic conflict in World War II, the political conflict in the inter-war Yugoslav monarchy, and the historical pattern of control by outside great powers.

This paper examines those factors in the current conflict in two ways. First, the course of those previous conflicts explains, at least partially, the pre-existing pattern of ethnic settlement and political boundaries. Secondly, the pre-existing pattern of ethnic settlement, political boundaries, and physical geography explain to a great extent the imbalance between the parties to the current conflict, and the strategies available to them.

The first part of this paper examines the roots of the ethnic and political geography which has led to the current conflict. The ethnic patterns in former Yugoslavia are primarily the result of previous conflicts between ethnic groups within Yugoslavia, previous conflicts between Yugoslav ethnic groups and outside great powers, and conflicts between outside great powers. The conflict between outside great powers seems to be the primary factor in the large-scale ethnic distribution. The political boundaries can also primarily be explained as a result of conflicts primarily involving outside powers, despite the periods of Yugoslav political independence after World War I.

The second part of this paper explains the strategy of the various sides in the current conflict. The imbalances between the three main parties to the conflict, Bosnian Serbs, Croats, and Muslims, are examined. These consist of an imbalance in military potential due to the sizes of the groups, differential access to safe rear areas and sanctuaries, and differential access to modern military equipment.

The ethnic distribution of the people supporting the conflict can help explain the military strategies of the parties to the conflict. Yugoslav ethnic geography is characterized by ethnic exclaves and enclaves, and the conflict can be shown to have followed a pattern based on the locations of those enclaves and exclaves. The sides which possess local military superiority have pursued proactive strategies based on the establishment of links between exclaves and other exclaves, and between exclaves and the main bodies of their territory. The sides which are militarily inferior have been forced into a reactive strategy, since moves by the superior side to connect isolated exclaves result in the creation of exclaves from the perspective of the inferior side. Evidence of this is found in current newspaper accounts of the conflict.

The conflict is of course also constrained by the unusual physical geography of former Yugoslavia. Thus, the fighting in Yugoslavia has focused on these patterns of demographic, political, ethnic, and physical geography.

Jim Simmons
University of Toronto
Toronto, Ontario Canada

COMPARING COMMERCIAL STRUCTURE: HONG KONG AND NORTH AMERICA

Retail chains are products of the environment in which they originate: including markets, retail structure, competition and government regulation. Their potential for expansion into new retail environments depends on their fit with these new locations.

While regional and national chains have easy access to market research that provides data to compare existing and potential locations, it is much more difficult to evaluate international markets systematically. The Centre for the Study of Commercial Activity is examining the general problem of comparing retail environments, and this paper uses the examples of Canada (Toronto), the United States (Los Angeles), and Hong Kong.

The results of this exercise emphasize (1) the diversity of measures required to evaluate new retail environments, including measures taken for granted in domestic studies, (2) the observed variation in these measures, which again lies outside the domestic experience, (3) the rapidity of change that can occur as explosive growth in market income generates new retail forms in new locations, and (4) the indirect effects of government regulation as it shapes competition.

Tim Sly
School of Environmental Health
Ryerson Polytechnic University
Toronto, Ontario M5B 2K3

ENVIRONMENTAL INFLUENCE ON DETERMINANTS OF DISEASE AND ILL-HEALTH

The environment has been recognized as a determinant of disease since the time of Hippocrates, more than two thousand years before the aetiologic agents themselves were identified and understood. Recently-identified changes in the environment, particularly in the atmosphere, forewarn of shifts in the range of certain disease entities, and also suggest widespread implications for the health of human populations through several physical mechanisms. The geographic distribution of communicable diseases may shift due to changes in temperature and other environmental conditions which effect the survivability of the agent and its vector or host, or alter the ability to transfer infection. Certain parasites and arthropods or public health significance already appear to be changing their distribution.

Physical characteristics of the environment having direct and indirect impacts upon the population include increased risk of malignant melanoma, the effect of ultra-violet radiation upon the immune system and specific target organs, and the temperature-related stress. Environmental stress generally is an area in need of intense study, as urban populations find themselves increasingly expected to deal with environmental threats to their health, perceived or real. Atmospheric effects upon food-growing zones, can be expected to manifest in global food and water shortages, reduced nutrition, and deteriorated health status.

Joel Sobel
Geography Division
U.S. Bureau of the Census
Washington, DC

USING LOCALLY MAINTAINED DIGITAL GEOGRAPHIC FILES TO UPDATE THE TIGER DATA BASE

During the 1980s, the U. S. Bureau of the Census dramatically transformed its geographic operations. It moved from traditional, mostly manual processes using paper maps and sequential record computer files, to state-of-the-art, interactive activities for creating and maintaining the TIGER data base. For several reasons, however, the Census Bureau did not design its geographic programs so that local officials and other participants also could take advantage of new computer and data base technologies. Instead, these program participants still delineated geographic areas, updated feature networks, and supplied address ranges by hand-annotating maps and listings provided by the Census Bureau. As the Geography Division begins planning its operations leading to the 2000 Census, its goal is to expand its geographic programs to include locally maintained digital geographic files in addition to the traditional annotated paper maps and listings.

The Geography Division refers to this process as Digital File Interchange. Many program participants already have informed the Census Bureau that their agencies regularly maintain digital geographic data bases. Just as the Census Bureau developed the TIGER System to integrate its geographic operations and minimize redundant, manual work, these local agencies now wish to eliminate labor intensive tasks to reproduce information they already have in digital form. However, for the Census Bureau to utilize externally-prepared data bases, it first must determine program requirements and standards. Among the key questions it must answer are:

(1) Which data base or GIS formats can it accept from program participants?

(2) What technical obstacles can the Census Bureau expect to encounter in attempting to integrate locally maintained digital geographic files into the TIGER data base?

(3) Which data items will the Census Bureau accept in digital form from program participants?

(4) What content standards must these data items meet?

(5) What accuracy standards must these data items meet?

(6) Must all data items be linked to TIGER records or to geographic areas recognized by the Census Bureau?

The Census Bureau now is working with a number of agencies at the regional, state, and local levels to answer these questions. For the 2000 Census, the Census Bureau expects that digital file interchange will be an essential component of its programs to acquire accurate, up-to-date geographic information.

Lawrence M. Sommers and
Assefa Mehretu
Department of Geography
Michigan State University
East Lansing, MI 48824

TOWARD AN EXPLANATION OF THE SOCIO-ECONOMIC DIMENSIONS OF MICRO-MARGINALITY IN SOUTH-EASTERN MICHIGAN

It has been observed, in the process of regional development, that two types of margins are revealed as a consequence of unequal and/or inequitable distribution of benefits. The more familiar variety is the macro-spatial margin; revealed over space as a horizontal distance decay function of any mean indicator of growth and development such as income per capita. The second type is the micro-spatial margin; which is less recognizable in macro-spatial
disaggregation. Unlike the macro-spatial variety, micro-spatial margins are less a function of distance-decay of development opportunities or of the imperfection of market forces but more a product of social, cultural and political factors such as class, ethnicity, nationality, gender and age. The aim of this paper is to explain the morphology and process of micro-spatial marginality in terms of the above mentioned factors and to measure their relative significance in contributing to historical growth of mirco-margins located within more prosperous areas.

Leigh Sparks
Institute for Retail Studies
University of Stirling
Stirling, Scotland, United Kingdom

BUSINESS EDUCATION LINKS: TEN YEARS OF THE INSTITUTE FOR RETAIL STUDIES AT THE UNIVERSITY OF STIRLING

The Institute for Retail Studies (IRS) at the University of Stirling is ten years old. The aims of the IRS are:

- to act as a centre for and to carry out academic and applied research into the retail industry;
- to provide a resource base for academic and industry-based researchers;
- to be involved directly in teaching students about the operation of retailing and processes within the industry;
- to provide management development services allowing application of research in universities to professional management in the retail industry;
- to create links and contacts with similar institutes in other countries;
- to foster and extend links between the University sector and the retail industry.

These aims are met by the full-time and visiting staff of the IRS pursuing their professional activities in education and research and reporting these activities through their publications. In the ten years since establishment, the IRS has become known internationally for its retail research and education. This paper reflects on the experiences of this period with particular emphasis on the industry-education links forged during these activities. This paper will comment on:

- the structure of the IRS;
- the industry-education relationships;
- the constraints on activities;
- the opportunities for the next ten years;
- the lessons (hopefully) learned.

The aim is not to provide a recipe for copying, but rather to look at a potential approach to the furthering of retail research and education. A variety of approaches are valid and much can be learned from the swapping of experience and discussion of the opportunities for industry and education.

Leigh Sparks
Professor of Retail Studies
Institute for Retail Studies
University of Stirling
Stirling FK9 4LA
Scotland UK

SPATIAL RESTRUCTURING OF FOOD RETAILING IN EUROPE*

The retail food market is the most important retail sector in volume and value terms in any developed economy. The largest retail companies in Europe are found in the food sector. Although the major product market, the food and grocery market in Europe is relatively static in volume terms as the proportion spent on food is falling across Europe, most notably in northern Europe. On the supply side the dominant trend is one of market concentration, although again there is a north-south divide.

These demand and supply side forces have essentially resulted in a market that is static, served by fewer organizations through larger stores. In this situation, a clear strategic problem arises for the larger companies when the cost of acquiring or developing further market share becomes prohibitive - in either financial or legal terms. Strategic responses have been to develop new retail markets, either geographically or via internationalization or sectorially via diversification.

Given this scenario, retailers have attempted to maintain growth within the grocery market via

- increasing the value of the grocery market;
- improvements in operational efficiency; and
- differentiating themselves from competitors to create customer loyalty.

Although these trends are common across Europe, the national or regional context will determine the direction and success of the marketing oriented responses to the changing grocery market. National and regional differences affect products and brands at one level, but also the nature of competition at another level. The legislative, historical and competitive differences influence the retail concepts and formats that develop.

This paper examines these topics with the aim of helping to understand and account for the restructuring of food retailing across Europe. The topics above are examined through illustrations of trends from amongst all European countries. The increasing inter-relationships and internationalization (the spatial restructuring) are placed in the context of retail development generally and in the specific factors discussed above. Whilst Europe is the focus, the paper concludes by considering Great Britain in particular and the opportunities and change therein.

---

*Europe here is taken to mean Western Europe

J. H. Springer
School of Urban and Regional Planning
Ryerson Polytechnic University
Toronto, Ontario

GROWTH MANAGEMENT AND HOUSING AFFORDABILITY IN ONTARIO

Growth management may be broadly defined as: "the imposition of guidelines to direct needed development to selected sites, while protecting an area's natural resources." (JAPA, 1992) When this broad definition is applied to Ontario, it is clear that the Province has in place a collection of regulations and regulatory agencies that together meet or exceed what is necessary to be considered growth management. However, in Ontario, growth management is subsumed within the generic term 'land use planning'.

This paper argues that while a strong case may be made for a broad based land use planning approach to growth management issues, the failure to recognize and address the special requirements that the growth management issues impose on the implementation and administration of land use policy result in a structure that Hamilton describes as, "lacking in adequate accountability, co-ordination and enforcement of existing regulations." (C.M.H.C., Symposium on Housing Young Families Affordably, 1990)

Judith Innes argues that, "Because growth management is so complex and involves so many actors, actions, and places, no one set of experts can design a successful program, nor can any state impose an effective program from the top down. Instead, groups including experts, citizens and high level officials, go through a process of mutual learning to create a shared conception of the intent of growth management and agree on specific ways to implement it." (JAPA, 1992)

A comparison of the mandate and focus of the Sewell Commission on Planning and Development Reform in Ontario, with that of the U.S. Advisory Commission on Regulatory Barriers to Affordable Housing provides a useful context in which the differences in focus and outcome can be discussed.

The study asserts that the growth management focus prevalent in the U.S. produces policy responses that address the specific impacts of growth management on housing affordability directly as reflected in the Advisory Commission's Report and in the N.J. State Planning Act of 1986 (which was an attempt to integrate the affordable housing issue in N.J. State Planning Act of 1986 (which was an attempt to integrate the affordable housing issue of New Jersey, with land use, transportation, economic development and environmental protection). In Ontario, however, the more general focus allow the questions of housing affordability to be addressed by a separate set of policies and structures. It is this approach that raises questions about accountability and co-ordination, especially with respect to affordable housing issues.

Carl A. Swanson
On-Site Inspection Agency
Washington, D.C.

THE BREAKUP OF THE NUCLEAR WEAPONS COMPLEX OF THE FORMER SOVIET UNION AND ITS IMPACT ON POST-COLD WAR ARMS CONTROL AND NUCLEAR PROLIFERATION

Concomitant with the dissolution of the Soviet Union in December 1991, was the breakup of its nuclear weapons complex. This nuclear weapons complex, consisting of deployed nuclear weapons and various components of the nuclear weapons production infrastructure, is now geographically divided among the four "nuclear successor states" to the former Soviet Union: Russia, Ukraine, Belarus, and Kazakhstan. This came at a time when the U.S. and the Soviet Union were making great strides in the negotiated "build-down" of these nuclear arsenals through the implementation of the Intermediate and Short Range Nuclear Forces Treaty (INF) and the signing of the Strategic Arms Reduction Treaty (START).

Now, with the creation of these three "de facto" nuclear states, in addition to Russia, the strategic nuclear balance of the bipolar, Cold War period has been upset. The relative stability of the U.S.- Soviet nuclear relationship has been replaced by the uncertainty as to how these new nuclear states will act with regard to these weapons. Although all have agreed to adhere to the Nuclear Weapons Non-Proliferation Treaty (NPT) and the START Treaty, Ukraine, in particular, appears to be having second thoughts.

Of further concern than just the positioning of deployed nuclear weapons systems on the territories of these states, is that, during this increasing period of political strife and economic hardships in these states, control of these weapons could be lost; existing weapons and/or their components, to include nuclear warheads, could be sold to third world countries seeking a nuclear capability; or that elements of the nuclear weapons production complex, located in these states, could be used to preclude nuclear weapons components, including weapons-grade fissionable, for hard-currency export. This would significantly undermine the progress made in limiting the proliferation of nuclear weapons under NPT.

This study is intended to provide a geographic overview of the nuclear situation (both deployed weapons and production capability) in the successor states to the former Soviet Union; analyze the current position of the governments of these new states to the weapons on their territories (especially Ukraine) and their responsibility to the arms control treaties signed by the Soviet Union to limit and reduce these weapons; discuss the implications for nuclear weapons proliferation; and finally look at some U.S. policy options to manage the controlled and continued "build-down of these weapons.

S. Martin Taylor
Department of Geography
McMaster University
Hamilton, Ontario, L8S 4K1

PSYCHOSOCIAL EFFECTS OF ENVIRONMENTAL CONTAMINATION

Previous research on the adverse human health effects of acute or chronic exposure to environmental contamination has emphasized physical health outcomes (e.g., cancer, respiratory disease, reproductive abnormalities). In recent studies, increasing attention has been paid to psychosocial effects. These are defined as the complex of distress, dysfunction and disability manifested in a wide range of psychological, social and behavioral impacts in individuals , groups and communities as the consequence of actual or perceived environmental contamination. Over the past five years, a multidisciplinary team at McMaster has conducted several studies of psychological effects in populations exposed to waste disposal facilities. This paper provides a review of the findings of this work and concentrates on the most recent study which has investigated psychological effects in the population exposed to the PCB contaminated site in Smithville, Ontario. In addition to describing the main results, the paper will discuss the contributions of the research program to the development of theory and method and the applications to on-going debates about the siting and management of waste disposal facilities.

Linda A. Thompson
AHCCCS
Office of the Medical Director

Susan McGraw-Helms
AZ Dept. of Health Services
Office of Women's & Children's Health

A NEIGHBORHOOD OUTREACH PROGRAM: DEMOGRAPHIC FACTORS AFFECTING SITE SELECTION AND ADAPTION

This paper seeks to examine the demographic factors that were considered in the site selections for a health promotion project in Arizona. UN COMIENZO SANO/HEALTH START is a Lay Health Worker Outreach Program designed to address barriers to prenatal care from the perspective of the women and families at greatest risk of poor birth outcomes. Lay health workers reflecting the characteristics of the community promote prenatal care and primary health care for children among low-income pregnant women and their families at the neighborhood level.

The methodology for site selection varies with the evolution of implementation in Arizona. The original sites were selected circumstantially. The following sites were selected by a Request for Proposal (RFP) process. The RFP specified the nature services, the model to be used, and the preferred target populations.

Specific demographic risk factors that have been associated with insufficient prenatal care constitute the selection criteria and were combined with qualitative data documenting the need of each of the three successful respondents. Social and culture features may categorize an area as high risk even where there is not a geographical barrier to the care. These traits may include minority status, age, education, birth order, marital status and poverty. Geographical accessibility is also clearly documented to impact utilization of prenatal care.

Successful respondents used a variety of outcome and utilization measures to demonstrate need in their areas. Analysis of each site according to the criteria described in the methodology demonstrates the appropriateness of the sites selected. This is true even of the original two sites, although these locations were not evaluated by the same process as those selected through the RFP.

Each of the five UN COMIENZO SANO/HEALTH START sites is clearly able to demonstrate geographical inaccessibility to care, high incidence of inadequate prenatal care, and several socio-demographic risk factors associated with inadequate prenatal care. UN COMIENZO SANO/HEALTH START serves at-risk population clusters where 50% to 70% of the population qualifies for Medicaid.

The individual RFPs defined their target neighborhoods. All sites used units or boundaries different from one another for this process. Defining the "neighborhood" to be served became even more complex during implementation. Neighborhoods with geographical boundaries became too limiting. A concept of "neighborhood," evolved to include the personal network of the outreach worker.

The next step to truly test the appropriateness of the site selection criteria would be a complete and thorough evaluation, measuring the numbers served, incidence of true case-finding, changes in utilization of prenatal care, birth outcomes and establishment of primary care for children. Refining the data collection process is necessary in order to analyze the effectiveness of each site. The Arizona Department of Health Services is currently seeking expertise to consider this need.

Nancy Torrieri
Geography Division
U.S. Bureau of the Census
Washington, DC

THE POPULATION DENSITY OF URBANIZED AREAS

A recent article in American Demographics identified twenty 1990 urbanized areas (UAs) recognized by the Census Bureau as having overall population densities of less than 1,000 persons per square mile (p.p.s.m.). This is significant because official Census Bureau criteria require that census blocks included within a UA meet or exceed this 1,000 p.p.s.m. population density requirement.

The explanation for the disparity between the overall population densities of some UAs and the official requirement of a minimum 1,000 p.p.s.m. population density lies in the UA criteria and the implementation of these criteria during the UA delineation operation. First, the UA criteria permit the designation of entire places as urban, including those blocks that do not meet the 1,000 p.p.s.m. requirement. Second, the UA criteria allow the inclusion of uninhabited non-residential urban land use areas and "undevelopable" land area under certain restricted conditions. In recognition of the complexity of settlement patterns around large places, the UA criteria also allow the inclusion of discontiguous high density settlement areas along with the low density "corridor" areas that connect these discontiguous settlement areas to the densely settled UA core. Finally, during the UA delineation process, the Census Bureau occasionally groups blocks with densities under 1,000 p.p.s.m. with other blocks having population densities that exceed 1,000 p.p.s.m. when the blocks are similar in street pattern and residential development. The overall population density of the residential portions of such block combinations always meet or exceed 1,000 p.p.s.m., but the grouping of blocks has some effect on reducing the overall population density of UAs.

By revising the UA criteria or devising new methods for UA delineation, the Census Bureau could ensure that the overall population density of all future UAs meets or exceeds 1,000 p.p.s.m. Such changes, however, may not be desirable. In addition, the implementation of such changes would require the support of those for whom the UA is a statistical basis in their research or policy implementation. The Census Bureau plans to continue its exploration of the most effective methods for UA delineation, and for devising the best possible measures of the nation's urban and rural components.

Gordon Wilcox
University of Windsor
Windsor, Ontario

AN INVESTIGATION INTO THE LOCATION OF THE FREEPORT CHRONIC-CARE HOSPITAL IN KITCHENER, ONTARIO

In 1987, construction began on the 350 bed Freeport chronic-care hospital in Kitchener, Ontario. Following the recommendation of the local health committee, and approval of the provincial government, Freeport became one of the only 14 chronic-care hospitals, in Ontario, with over 200 beds. The purpose of this investigation is to determine whether Kitchener was the most suitable location for Freeport, and if not, where was the hospital's most favourable location.

The first consideration of the study was to determine those sites which may have been optional for the construction of the hospital. Accessibility and utilization patterns were collected for all sites in order to compare the suitability of all the optional sites within the province. The attribute information for each of the optional sites was compiled for comparison within a computer-based decision support system (DSS), known as DINAS (Dynamic Interaction Network Analysis System). DINAS is designed to compare specified attribute information about the optional sites, in order to produce a ranking of the sites, from best to worst.

The study itself questions the system of hospital allocation within Ontario's public health care system. Presently the provincial government allocates funds for hospital construction based upon the recommendations of local health committees, and approval of the Ministry of Health. Furthermore, the study attempts to contribute to the growing literature concerning the use of computer-based decision support systems.

Nancy L. Winter
Graduate School of Geography
Clark University
950 Main Street
Worcester, MA 01610

HAZARD MANAGEMENT MAPPING: A TAXONOMY

Hazard research and cartography, disparate geographic endeavors, set the interdisciplinary context for hazard management mapping. Natural and technological hazards occur over a variety of spatial dimensions. Cartographic means for depicting the spatial patterns of the locations of hazards, their degrees of risk, methods for hazard control and response, and hazard mitigation strategies have the potential to play a unique supporting role in hazard management. Mapping for hazard management ensues in both theoretical hazard research and in applied situations. Integration of cartographic wisdom with the findings of hazard research is essential. In particular, the distinction between the terms "hazard" and "risk" must be recognized. Hazards involve the identification and location of threats to humans and what they value while risks require a quantifiable measurement as to the degree of threat from a single hazard or multi-hazards. The two terms should not be used interchangeable.

Maps are too often a neglected tool in the practical handling of hazards. Three basic ongoing processes are involved: 1) map design and production, 2) map use for planning, crisis control, risk communication, and pre-post mitigation attempts, and 3) map influence, affecting risk perception and perhaps even political decisions. Application of maps to solve hazard problems demands a taxonomy depicting the functions executed with hazard management maps and furnishing clear definitions for map types. Four main objectives of hazard management characterize the roles for maps.

1) HAZARD IDENTIFICATION requires *hazard maps* which identify and locate hazard zones. These often portray Donald Deigler's "hazard geometries."
2) RISK ASSESSMENT AND SOCIAL EVALUATION dictate a need for *risk maps* which show results of calculations of the degree of threat to humans and what they value and depict the spatial distribution of the risk assessment numbers.
3) EMERGENCY MANAGEMENT requires *emergency maps* to cover the four aspects of emergency management charted by Perry, Lindell, and Greene in 1981 - mitigation, preparedness, response, and recovery - with five potential purposes:
    - to serve as the quickest method for locating at a glance all the elements at work in a specific geographic area (without having to read large volumes of information)
    - to coordinate the efforts of emergency groups through use of a generally acceptable and concrete model of the area impacted by a hazardous event and by depicting key elements of emergency planning

- to aid in control of the physical agent whenever possible
- to aid in the flow of resources and services
- to provide a guide for possible action by the public

Two major categories of emergency management maps exist: *emergency planning maps* and *crisis maps*. Emergency planning maps, generally prepared before a hazardous event takes place, include special content maps such as *evacuation maps* which identify sensitive populations and may provide a visual plan for the public to use to evacuate before, during or after a disaster event. *Crisis maps* are produced by emergency managers on-the-spot during a disaster and range from simple hand drawn sketches on cardboard to sophisticated GIS computer maps.

4) In RISK COMMUNICATION, the three types serve as educational devised or public relations tools and may affect risk perceptions.

Disasters precipitated by natural or technological hazards require the movement of people, information, and goods and services over space. For the most effective hazard management, the production of hazard maps, risk maps and emergency management maps, need to be fully funded before, during, and after a natural or technological disaster. As threats to humans and what they value are perceived to exist on an ever greater global scale, the process of mapping should be carried out at all stages of hazard management and all levels of human involvement.